Why You Need To Read This Book

People around the world dream of mathematical excellence. Yet, so few will ever develop the math skills they'll need to succeed.

As a result, they may never pursue the amazing hoped for job opportunities. They may never write computer programs or take part in the creation of the kinds of formulas that send spaceships to the moon. They may never learn to balance their savings and expenses, which may mean a life of uncertainty, stress and debt. They may never have impressive business contacts or travel the world, all due to failure in this area.

Even with the best intentions and instructors, students struggle to become "fluent" in mathematics.

Why is math such a struggle? Many math students blame a lack of time. Some claim that memorizing the math rules and formulas they need to know is too hard. Some claim they weren't born with a "math brain," are mathematically "challenged," or blame heredity – "No one in my family has ever understood math." Others try to learn math concepts by rote. They copy formulas hundreds of times into their notebooks by hand, praying that the information will stick.

The biggest excuse heard around the world, and in every language, is the saddest excuse of all. Most people who struggle with math claim they have a bad memory.

I sympathize with this. I used to love claiming, "I have a poor memory!" In fact, I have silently sworn so vehemently about my "bad memory" that, had I spoken my frustration out loud, my teachers would have kicked me out of class.

I remained irritated with what I perceived to be my poor memory until I decided to do something about it. I studied memorization and ultimately devised the unique Memory Palace method described in this book: The Magnetic Memory Method.

What is the Magnetic Memory Method? It is an easily learned set of skills that you can completely understand in under an hour. It is a method that will have you memorizing numbers and math rules at an accelerated pace. Within just a short few hours, after you've learned the technique, you'll amaze yourself by what you can do.

Instead of struggling to learn and retain one or two formulas a day, you will find yourself memorizing much, much more. Every time you learn math in conjunction with the memory techniques taught to you in this book, you will gain more knowledge of math. Your imagination will also get stronger, which leads to a whole host of other benefits.

It pleases me immensely to help people memorize numbers and mathematical equations. I'm always delighted when people write to me with stories of success. I receive thank you notes almost every day as readers of my books describe to me how easily they memorized sometimes very difficult information.

These achievements are thrilling to me, thrilling for the people who use the techniques and they will thrill you also.

This edition of How To Memorize Numbers, Equations & Simple Arithmetic is for you. Whether you are an adult, teenager or someone working with young students, you will benefit from this book.

Anyone who struggles with learning, retaining and using numbers and formulas stands to gain a great deal. I have designed this book so that as soon as you understand the core memory method, you can sit with a math textbook and memorize. Anywhere and at any time you can recall any rule that you wish easily and accurately. Permanently.

I have written this book for those math students who have the burning desire to learn a formula once and recall it within minutes, if not seconds, of having learned it, and to do so without frustration of any kind.

Three obstacles stand between you and memorizing numbers, formulas and arithmetic. The belief that:

1. You don't need a dedicated memorization strategy for memorizing numbers and formulas.

2. Memorization strategies won't work for you.

3. Memorization strategies are too much work.

Let's examine each of these beliefs.

The Belief That You Don't Need A Dedicated Memorization Strategy For Memorizing Numbers and Formulas

Although repetition is always important when it comes to learning, it is a shame that so many people wind up relying on rote learning. I call learning by rote the "blunt force hammer" of education. Why? Because it is exactly like pounding your eyes and brain with a hammer. This is especially painful to see when there are ways to use your natural imagination, ways which expand your ability to learn and memorize as you memorize and learn.

Worse, people who use rote learning are usually deluding themselves. We have all experienced the fantasy that repeatedly looking at index cards will put the information into long-term memory. However, only rarely does this painful activity reward you with permanent ownership of those numbers and formulas.

The fact is that repetition without making a memorable connection with the material doesn't work – not in my experience, and not in the experience of the thousands of people who have read my Magnetic Memory Method books and taken my video courses.

Looking back at my own student experiences, I'm shocked that my schools did not teach dedicated memorization skills. Instead of sitting through long classes based upon the repetition of one or two math exercises, I could have been using an alternative. With a dedicated memorization strategy, I could have been memorizing dozens of important math concepts per day.

The Belief That Memorization Strategies Won't Work For You

People often tell me that the memory techniques I teach will not work for them. I always confidently respond with a simple truth:

Not only will these techniques work if you follow the exercises, these techniques will literally blow you away.

Especially when you see how quickly your math skills develop.

Try out the memory techniques taught in this book for yourself and you will marvel at the progress you'll make. Guaranteed.

The Belief That Memorization Strategies Are Too Much Work

You will need between 1-2 hours to set up the full Memory Palace system taught in this book and another 2-3 hours to really get the hang of the method. After that, it's just a matter of picking up speed. The steps are easy and fun. The bonus is you can memorize numbers and formulas as you are learning the Magnetic Memory Method.

As soon as you've understood the principles of math memorization and have started working with the method taught in this book, you will be memorizing new math rules by the dozens – all with consistent speed and accuracy. The best part is that this method will serve you for life and can be extended to memorizing just about any information you could ever want.

I have a suggestion for you before you turn the page and start your journey toward advanced memorization skills. Believe in the power of your mind.

When I started using memory techniques, for example, I constantly told myself that the language, or subject matter I was learning at the time, was too difficult and that my brain was ill equipped. I acted as if I had been born with a poor memory. This not only pushed the information to be learned away from me, it also eroded my confidence and made things much more difficult than they needed to be.

Don't be like this.

The ability to memorize numbers and math rules and put them to use with near-100% accuracy opened the world's doors for me, and it will do the same for you.

Moreover, when we consider the importance of math in society, it is that much more important that we do not belittle ourselves. Your mind is powerful. By developing a positive mental attitude and learning the Magnetic Memory Method your efforts will be easy, fun and demonstrate to you the powerful abilities of your own mind every single day for the rest of your life.

Math remains a "language" spoken all around the world. This means that those with solid math skills can experience so much more in their careers.

With advanced math skills, you'll qualify for better jobs. You'll have more opportunities in the fields of science, computing and engineering. If you are a businessperson, you will engage in meetings and meet potential clients and partners with the ease and efficiency of number mastery that marks all great entrepreneurs. And, there are many more benefits too.

You will love adapting the Magnetic Memory Method for memorizing numbers and formulas to your individual learning style, and you'll enjoy massive success as a result.

Give me 5 hours of your time (or less) as you teach yourself how to use this method. In return, I will give you the techniques and abilities you'll need to memorize all the math formulas you have ever dreamed possible and experience massive boosts in number and math "fluency" as a result.

How To Learn & Memorize Math, Numbers, Equations, & Simple Arithmetic

By Anthony Metivier, PhD
www.magneticmemorymethod.com

Wait!

I have created FREE Magnetic Memory Method Worksheets just for you. These worksheets will help you take the memory improvement lessons you'll learn in this book to the next level. You'll also be given the opportunity to watch the free video course *Memory Palace Mastery.*

In order to download these worksheets and start watching the videos, go now while these materials are still FREE:

http://www.magneticmemorymethod.com/free-magnetic-memory-worksheets/

Table of Contents

Introduction

First off, I want to congratulate you on laying a new foundation for your math learning experience. This book is truly groundbreaking. For the first time there is a collected package of tools, strategies and insights needed to succeed with memorizing mathematical principles and formulas.

Therefore, the opportunity you have before you now is indeed an exciting one. You now have in your possession the same information and material that has enabled thousands of ordinary men and women with no special memory abilities to raise their passion for math to the next level and experience massive boosts in their (and now your) understanding of and ability to understand and use math.

Why is this book so powerful?

Because nothing about building and using Memory Palaces for math has been held back, you're getting everything there is to know about using Memory Palaces.

This introduction to the Magnetic Memory Method includes the most complete and detailed training on building a network of Memory Palaces for memorizing math ever presented. The best part is that you can use the technique to memorize any other kind of information you'll ever encounter.

In addition to this amazing training, you'll find:

* A complete description on how to build and use Memory Palaces for memorizing and recalling mathematical formulas, principles and rules.

* Access to the author to answer all your questions (my email address is learnandmemorize@zoho.com).

* Secret strategies for using relaxation to aid the memorization process.

* Tips about overcoming procrastination while studying math.

* An exclusive, Preferred Reader invitation to receive ongoing content to back up the memory training offered in this book.

* Access to amazing bonuses linked to throughout this book that will inspire you and deepen your familiarity with using Memory Palaces.

As you can see, this is a VERY full book.

Now, you might be wondering, what do I do first?

Here's a STRONG recommendation. Take a moment to answer the three quick questions on the next page. You can easily email me your answers with "Memory Questions Answered" in the subject line for a free gift that will continue your education in the art of learning and memorizing.

That's it for now. You have lots to do and a very exciting adventure ahead of you! Make sure you subscribe to the Magnetic Memory Method newsletter and watch your email inbox for ongoing announcements, and make sure to get in touch with any questions you may have by email at learnandmemorize@zoho.com.

Dedicated to improving your math knowledge and your memory,

Anthony Metivier
Founder of the Magnetic Memory Method
www.magneticmemorymethod.com
learnandmemorize@zoho.com

Send In Your Answers To These Three Questions For A Special Gift!

Email me your answers with "Memory Questions Answered" in the subject line at learnandmemorize@zoho.com.

Magnetic Memory Question #1:

What is your personal "Memory Myth" about your memory, including any programming you may have received as a young person or continue to receive in your daily life? How does this myth affect how you think about your memory?

Magnetic Memory Question #2

What is the "distance" between where you are now with your memory skills and where would you like to be in the future? Please be as specific as possible, including something like a deadline for when you would like to see a difference achieved (five minutes from now, tomorrow, next month, next year, etc.).

Magnetic Memory Question #3

What is your education "action plan" for completing this course so that you have total control over the improvement you would like to see in this area of your life?

Remember: email your answers to me with "Memory Questions Answered" in the subject line at learnandmemorize@zoho.com for a special gift ($24 value).

Chapter One: The Basics Of Number Memorization

The ability to remember numbers has many benefits. You can, for instance, memorize and recall such important sequences as:

* Social Security numbers for everyone in your family

* Your driver's license number

* Credit card numbers

* Birthdays and ages

* Coordinates

* Street addresses

* Phone numbers (Yes, the ability to memorize these still comes in handy. You never know when you're going to make a hot date at the swimming pool with no cell phone in sight!)

The good news is that there are specific mnemonic techniques for memorizing numbers. They're easy to learn and easy to use.

The typical go-to method is the Major Method. It's not called "Major" because everyone uses it, but because its fame is often attributed to Major Beniowski. We now know that an earlier version of the system already existed, invented by the French scholar Aime Paris. Paris, renowned for his number memorization techniques in the early 1800s, earned the honorable title of "professeur de mnemonique" from the Athenee University in Paris.

Other terms for Paris' method are the "phonetic mnemonic system" and the "digit-consonant" system. No matter what you call it, the basics of the Major Method consist of linking numbers with sounds. There are complex renditions of the Major Method, but the simplest goes like this:

0 = s
1 = d, t
2 = n
3 = m
4 = r
5 = l
6 = ch, j or sh
7 = k
8 = f or v
9 = p

Putting the sounds together involves inserting a vowel. To memorize a simple number like "22," you could insert "u" to make the word "nun," or if you're familiar with Indian bread, you could use "nan."

To take a longer example, "animal" could help you recall the number "235" because n = 2, m=3 and l=5.

But is this enough?

Not really.

What we need is to take these images and make them large, bright, vibrant, strange, bursting with color and energized with action.

For example, imagine needing to memorize "22235."

We already have "nun" (22) and we already have "animal" (235), so let's add zany action by having the nun attack the lion.

If the number were 23522, you could just reverse the image. Now the lion is attacking the nun.

By making sure that the words we create from the phonetic sounds are linked to the numbers, we make everything much more memorable.

You can think of this action-based "associative-imagery" as a kind of mini-story or vignette. In the scientific literature, images like nuns attacking lions are sometimes called "story mnemonics."

Permanent or Flexible?

Some people like to choose permanent images for numbers. For example, 22 would always be a nun and 235 would always be an animal.

My preference is to keep a small pool of figures to fall back on but maintain flexibility when needed. It's also important that the words we create from the Major Method phonetics are concrete.

What does this mean?

Compare "nun" to "none." Which one can you see in your mind?

Having completed this exercise, it should be obvious that memorizing "none," i.e. nothing attacking a lion, will not produce a strong memory that is easy to recall.

What about creating images for single digits?

Memorizing single digits is as easy as creating images for 1-9 in a way that requires no "phonetics" as such. For example, people often associate:

1 with a candlestick ...

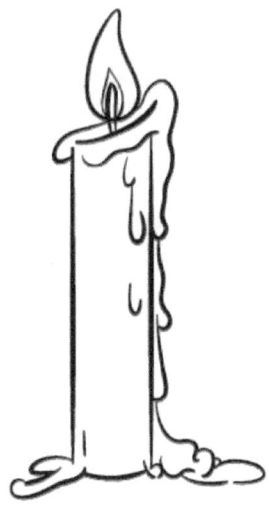

2 with a swan ...

3 with a sideways mustache ...

4 with a sailboat ...

8 with a snowman ...

... etc.

If you're going to create a "set and forget it" set of images that you use all the time, it's important to make sure that they come to you naturally. You should also put them into use right away so that your mind learns these "keywords" through use as well as memory.

And that raises a good point about mnemonics in general. It's not just that we link one thing with another for fun. We do it because these techniques are useful in ways that can make a real difference in our lives. That's why you need to use the techniques as soon as possible so that you not only understand how they work, but you feel how they work also.

It is difficult to express just how good it feels to be able to recall long strings of numbers with ease until you've done it. But, once you've done it, you'll have a hard time not "showing off" a little and teaching everyone you know how to use these new skills for themselves.

Grouping

One neat strategy involves different kinds of grouping numbers together. We've already talked about nuns attacking lions for grouping 2 and 3-digit numbers and you've seen how easy that can be. But what about a 7-digit phone number?

Anchoring figures is one solution.

For example, let's say that you encounter a phone number that starts with 2. The full number is 275-8923.

Assuming 2 is a swan in your single digit system, you can use it to "anchor" the rest of the numbers. Starting at the head, for example, you could see your friend Karl (k+l = 75) sliding down the swan's neck to insert a viper (v+p = 89) into a garden gnome's mouth where he is standing on the swan's back (n+m = 23).

Let's break this down again:

The swan is the anchor = 2

K+L (Karl) = 75

V+P (Viper) = 89

N+M (Gnomes Mouth) = 23

Notice that correct spelling goes out the window in this example. We're focusing on sounds alone using the Major Method. At the beginning, it can take practice to make these substitutions with speed. Yet, most people pick it up quickly.

You'll notice too that your mind has an amazing ability to hone in on exactly what you need to recall the number. The swan will come first, Karl sliding down the neck next, followed by the viper and the gnome. You may even find that you no longer need to mentally "see" these things. It will be more than enough to think about the associative-imagery to create the desired effect.

That desired effect of instantaneous recall is the closest thing to real magic we have. You can use this procedure to memorize almost anything. When it comes to numbers, you can link any number that starts with 4 to a sailboat, with 8 to a snowman and so on.

One thing you might be thinking is:

Hold on! I live in a city where everyone's number starts with 2. How many swans am I supposed to have floating around in my head?

The solution?

Read on, dear Memorizer, read on. We'll deal with this in detail in the next chapter. For now, let's read the ...

Chapter Conclusion

So far we've talked about unstructured approaches to remembering numbers and semi-structured grouping. To review: making single words like "animal" or "nun" amounts to an unstructured process. Putting them together by having the nun attack the animal is semi-structured.

We then added a bit more structure by using an anchoring figure, in this case a swan.

Structure is important because the more of it you have, the more you can let your mind fall back on it. In effect, building structures reduces, if not eliminates, "cognitive overload."

What is cognitive overload? It's the consequence of having so many things going on at once that your mind makes little or no progress.

But, what if with a bit of forethought and preparation, you could reduce, if not end, cognitive overload so that it never stops you from using your memory again? What if there were more structures we could use to give these powerful and memorable images a permanent "home" in our minds? Structures that are easy to find, easy to use and need almost zero effort to use?

If the answer to these questions interests you, then I am about to show you a fascinating technique. It is a technique that will let you hold as many numbers in your mind as you could ever wish.

Chapter Two: How To Build & Use A Memory Palace

In this chapter, you'll learn about the Memory Palace concept. We're going to get into a lot of detail about constructing well-formed Memory Palaces, but for now, sit back, relax and let the concept sink in. Memory Palaces will provide you with the ultimate organizational system, a cheat or crib sheet for your mind.

The best part is using Memory Palaces to store information in your mind is never cheating. Everything you've memorized has been learned in a legitimate way. You just learned it faster and more "magnetically" than anyone else did.

We've already talked about grouping. Memory Palaces take your number memory game one massive step further by "super-grouping." When you use all the techniques I've already described in this book, you are in effect making your mind "Magnetic."

As an important aside, let's look at this term "Magnetic," talk about why it deserves capitalization and what it means for you.

The Magnetic Memory Method uses the term magnetic for two reasons. First, it is about attracting information in a way that makes it stick in your mind for as long as you want. Second, using the other feature of a magnet, the Magnetic Memory Method helps you repel information.

Why would you want to repel information? Because there are a lot of things that you don't want in your memory which includes excessive information that causes cognitive overload. The Magnetic Memory Method and the Memory Palaces help you create and let you focus on the most important thing: getting only the information you want into your memory so that you can repel the rest.

For better or worse, it's difficult to explain exactly why this occurs. But once you're using the Magnetic Memory Method, you'll feel it.

What Is A Memory Palace?

A Memory Palace is a mental construct based on a familiar location. It allows for rapid and efficient journeys where you "meet" the associative-imagery you've created using the techniques in this book. Memory Palaces are the best way to store and recall information in a way that takes you to the level of memory expert. They create such massive success for Memorizers because they rely on actual locations.

Why Are Actual Locations So Important?

The answer is simple: The mind has the incredible ability to recall places that you already know with ease. By "places," we mean buildings in particular, rather than outdoors locations.

And this is why I suggest that you build Memory Palaces based on buildings. My experience, along with feedback from hundreds of my readers, demonstrates that buildings make for better Memory Palaces. This is because they come pre-structured.

Forest paths and beaches, on the other hand, come without structure and you need to impose artificial order on them. You need to impose order on buildings like your home too, but in a way based upon an existing architectural order. The patterns upon which you can "overlay" a reliable journey in a forest in no way compare to the stability of a familiar floor plan.

I do not want to confuse you with a contradiction here, but I would like to invite you to hear an alternate opinion. Phil Chambers, Chief Arbiter of the World Memory Championships, gave us an interview on the Magnetic Memory Method Podcast. In our discussion, Phil gives his opinion about indoor and outdoor locations as Memory Palaces. You can listen free here:

http://www.magneticmemorymethod.com/phil-chambers-talks-about-the-outer-limits-of-memory-skills/

Be sure to subscribe to the Magnetic Memory Method Newsletter while you're on the site! :)

In brief, Phil suggests that outdoor Memory Palaces pose no disadvantage compared to indoor locations. But we need to keep in mind that Phil is also a distinguished Memorizer with many years of experience. It's up to you to experiment with what gets results, but my advice remains the same. Indoor Memory Palaces come pre-structured. All you need to do is chart a journey through them.

Memory Palace History

Before we begin learning to build Memory Palaces, it will serve you to know a little bit about their history. No one really knows about whether or not the following story is true, especially given that there are so many variations of it to choose from, but as we'll see, what really matters is that the legend has clues about how to use memory techniques within it. I suspect it is for this reason that the "origin story" of Memory Palaces has survived.

Back in Ancient Greece, Simonides of Ceos (c. 546-468 BCE) found himself giving a speech at a banquet before a group of distinguished guests. The building collapsed and everyone but Simonides died.

In some versions of the story, Simonides was called out of the banquet by Castor and Pollux, mythical boxers who represent heroism. There doesn't appear to be any reason these two figures called him out of the banquet, but the occasion did save him from being crushed to death.

Regardless of how the story is told, because Simonides knew the secrets of combining images with locations, he knew exactly where everyone in the building had been sitting and was able to help the authorities identify the bodies so they could be properly buried by the mourning families who would never have experienced closure otherwise.

It is Simonides' ability to do this in combination with the building itself that led to the creation of the Memory Palace technique. The major point of the story that we will be referring to many times in this book is that Simonides used location to "store" and "revisit" memorized information.

The Important Mechanics of Memory Palaces

Location-based memorization is useful on many counts, mostly because it allows us to leverage the mind's natural ability to mentally organize space without significant effort.

Try this: close your eyes and visually reconstruct the room you're sitting in with your imagination. Chances are that you easily can do so. You might actually "see" it or only see a kind of floor plan made of simple shapes. You may even only "feel" or "sense" the concept of the room, but, one way or another you can reconstruct the room in your mind.

After that simple task is done, mentally move out into the hallway and reconstruct that space. Move throughout the entire building, recreating its rooms and its nooks and crannies in your mind. Work on making it visual, or simply develop what is now becoming a Memory Palace in whatever way works for you.

What we are doing in this exercise is using something we already know to create a powerful mental "link" that can be revisited at will with (almost) zero effort. We can place information at various points in this mental construct, "magnetize" that information by using associative-imagery and then revisit it later in order to retrieve the information we've memorized.

At least at the beginning, we want to always use what I call "non-arbitrary spaces." They are non-arbitrary because they mean something to us. These include places like:

Your home

Homes of relatives

Homes of friends

Libraries

Movie theatres

Hotels

Grocery stores

… and the list never really ends. You can always visit new places at just about any time that you wish in order to develop new Memory Palaces.

About The Term "Memory Palace"

Speaking of which, "Memory Palace" is the sexy term for "non-arbitrary space." Some people don't like the term

"Memory Palace," so if you're already gagging at the idea of using it as we'll be doing throughout this book, feel free to find a replacement. I once coached on 80 year old man through email who went on to memorize hundreds of lines of poetry using the Magnetic Memory Method, but only after he finally decided to call his Memory Palaces "apartments with compartments" because he found the term "Memory Palace" too hard to bear.

Whatever you do, don't let the terminology get in the way of making progress with the techniques you're going to learn in this book. Simply come up with your own if you don't like the terms. Maybe you'll wind up writing a book of your own one day and come up with something even more fashionable!

This raises the interesting questions of why we call them "Memory Palaces" in the first place. There are many potential answers, but one of my favorites appears in St. Augustine (354-430 ACE). In his Confessions he wrote "And I come to the fields and spacious palaces (praetoria memoriae) of my memory, where are the treasures of innumerable images, brought into it from things of all sorts perceived by the senses."

This is important because Augustine is pointing out the important fact that in order for Memory Palaces to become useful, we need to combine locations with all of our senses in order to create "treasure." By putting sensations together with locations, we can make information Magnetic so that it will come back to us whenever we wish.

You might also find it useful to know that location-based memory techniques appear to have existed before people like Augustine and Simonides worked with them. In her book on the Buddha, Karen Armstrong mentions the use of memory techniques in Yoga involving locations and the Buddhist instructor Michael Roach has spoken in great detail about how various meditations were remembered by the monks by placing imagery in different parts of the temple.

For example, in a meditation which asks us to remember that death is always behind us, monks were advised to place a black dog at a particular part of the temple to remind them of this principle. Interestingly, later religious traditions like Catholicism would take such ritualistic reminders out of the imagination and externalize them in the form of reliefs or paintings on the walls of their churches in the form of the Stations of the Cross.

The principle we learn from these practices is that we can divide Memory Palaces into "stations" that form the stops along a mental journey based upon "actual" journeys that you can actually take.

There are in fact two types of stations:

* Macro-stations

* Micro-stations

A macro-station is an entire room (i.e. bedroom, kitchen, living room, bathroom). A micro-station is an element inside of a room (i.e. a bookcase, bed, TV set).

It's important to recognize the difference because at the beginning stages of using Memory Palaces, it's often best for people to start out with macro-stations until they get the hang of the techniques. However, many people "get" how this works right away and that enables them to make quick progress right away with micro-stations.

The most important thing here is to get started ASAP. The sooner you start experimenting, the sooner you can start getting results from these amazing memory techniques.

Have a look at the following diagram.

Each of these rooms represents a macro-station. There are four in total, five if you count the area in front of the entrance.

Compare this with the same home, this time with labels that indicate just some of the possible micro-stations in this home:

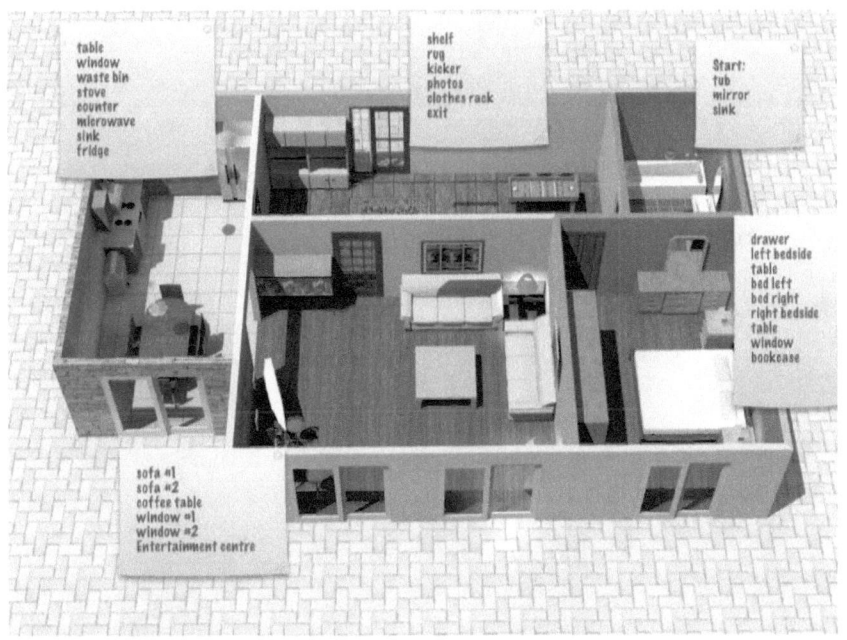

By identifying a liberal amount of micro-stations within each room, we now have 30 individual spots upon which to place information that we want to memorize. Thus, it's a great idea to work up to this level as quickly as possible so that you can take advantage of Memory Palace journeys that offer this many possibilities.

For the sake of completeness, here is a list of the stations indicated in the image above:

Tub

Mirror

Sink

Drawer

Left bedside table

Bed left

Bed right

Right bedside table

Window

Bookcase

Sofa #1

Sofa #2

Coffee table

Window #1

Window #2

Entertainment center

Table

Window

Waste bin

Stove

Counter

Microwave

Sink

Fridge

Shelf

Rug

Kicker table

Photos

Clothes rack

Door

Exercise

In order to let the power of building a Memory Palace journey sink in, take a moment to identify the Memory Palace macro and micro-stations in your own home. It doesn't matter if it's a house, an apartment building or a trailer. Even if you're reading this in prison, you can build a Memory Palace using your present location (I know this for a fact because both prisoners and prison guards have written to me to tell me about their experiences using them as Memory Palaces).

To fully benefit from this exercise, I suggest that you:

* Draw your Memory Palace by hand in a notebook

* Create a top down list

In other words, tap into both your visual imagination and your conceptual, organizational imagination.

As you construct your journey through the Memory Palace by identifying your stations, obey two key principles:

* Do not trap yourself

* Do not cross your own path

You want the journey you create to be linear because this makes it easier to follow the journey in your mind and you will spend much less mental energy when using the Memory Palace to store and recall information.

Moving from the visual example I've given you on the previous pages, let's look now at a real set of micro-stations in a real Memory Palace that I actually use.

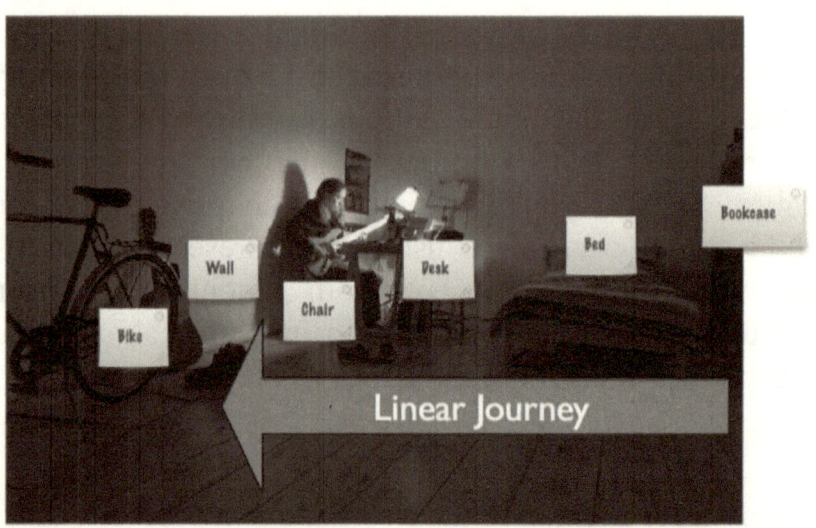

This is the office where I work. The bookcase stores books. I use the bed to study the effects of meditation on memory and research dreams. I use the desk and chair to write books and work on music memorization, the wall to lean my guitars on and the bike takes me home at the end of the day.

I also use all of these "micro-stations" to store information that I want to memorize. By making the journey linear with no path-crossing and moving towards a door so that I'm not trapping myself, there is no confusion about what comes next along the journey and mental energy expenditure is kept to a bare minimum.

Now that you've had a look into just one room of just one of my many Memory Palaces, are you beginning to see the power of separating places that you already know into individual stations so that you can use them to "drop" pieces of information in order to access them later?

I certainly hope so because there is literally no other memory system this powerful, and there is still so much to learn.

Specificity

In order to be truly successful when using Memory Palaces to store and retrieve large amounts of information, it's important that each Memory Palace is selected with care. Your Memory Palaces should be project specific. You want the Memory Palaces you use to respond to specific needs.

For example, I've started learning Japanese. To deal with the hiragana, I needed a Memory Palace with 48 stations that were tightly linked together, but not overwhelmingly so. After some thought, I drew a quick sketch of my girlfriend's apartment. Within five minutes, I had 48 stations written out in a list and 15 minutes later I had memorized both the sound and the shape of 15 characters.

It's really that simple.

However, if I had picked a Memory Palace that was too small or even too large and tried to work with it for this particular set of information, I can predict based on long experience with using Memory Palaces that my results would not be nearly as fast or as easy.

Thus, we should always work towards having an economy of means in our Memory Palaces. This term comes from the theatre and from film and refers to using the absolute bare minimum needed to express certain features of a story. A character who is depressed is often cramped by the camera to show isolation and despair. A character who is happy or free is given more space. Space is never wasted and has deep metaphorical value in most good movies.

In the world of Memory Palaces, too much space can lead to "decompression." We often want to pack our Memory Palaces tightly in order to maximize, not just the amount of information we can store in them, but also the energy.

How do you learn about this and get it right?

By building and using Memory Palaces.

I can only give you the guidelines.

Only you can undertake the journey and experiment with what works best.

Before moving on, another reason why you want to make your Memory Palaces specific to the information you're trying to memorize is because it helps you track your results.

For example, I told you that I memorized 15 hiragana characters in 15 minutes, knowledge that I could easily express because each station is counted. I could then predict how much time I would need for the remaining characters and budget my time accordingly. We're going to talk about this issue more in the chapter on using index cards in conjunction with Memory Palaces where you will learn how to memorize massive amounts of information in an even more structured way by deciding in advance how many stations you'll need and choosing your Memory Palaces for appropriateness before you even get started.

Chapter Summary

Choose your first Memory Palace by identifying a familiar location. Many people suggest that you should use your own home as a beginner, but I think you can be more adventurous if you wish. Use your school, church, workplace – nearly any indoors location will do, keeping in mind that you want to make it indoors to maximize the effectiveness and you want to be familiar with the location to the point that even without revisiting it, you can create a journey throughout the location in your mind and divide the journey into stations.

When working with your first Memory Palace, decide first whether you want to start with macro-stations or get right into using micro-stations. My preference is for people to start large with macro-stations and then narrow in to using micro-stations, but I leave this to you. Ideally, you'll try both, but there's only one first time and it's important not to frustrate yourself if you feel in advance that using micro-stations might be too much.

On the matter of overwhelm, make sure that you construct your first Memory Palace journey in a way that neither traps you nor enables you to cross your own path. This can be admittedly tricky in some buildings and may mean that you need to abandon features that you could otherwise use as stations in order to keep a linear journey that does not lead to crossing your own path or trapping yourself. But if you rely upon the principles I've given you in this chapter, your journey will be streamlined, easy to navigate and effective.

And if you're worried about not using all available space because you've left a number of micro-stations behind, don't worry about it. In the long run, it is always worth it to lose a few stations in favor of having journeys that are clear, linear and easily navigable. You do not want to lose mental energy, certainly not when you'll be using your Memory Palaces for the purposes of passing exams.

This is why it's important to focus on developing an economy of means, a tight and focused approach to getting what you want when you want it without having to remember anything about the journey you created. This concept of the economy of means works in film and it will work in your mind to create compelling journeys that help you recall all the information that you'll ever need to memorize.

When creating your Memory Palaces using these important principles, draw them out and keep a top-down record.

This means not only creating the floor plan so that you can see it visually, but also conceptualizing it logistically. By putting the two mental processes together, your mind will "solidify" each and every Memory Palace you create using this process, the beautiful result of taking a few extra seconds to let your brain interact with a location it already knows using more than one perception modality.

As a brief aside, although the image I've created above may not be sexy, that's the point. You don't need to be an artist or a graphic designer. You just need to link your mind with your hand in order to create a stronger link between what you will soon rely upon only in your imagination and the reality of that location in the real world.

This brief exercise will also help ensure that you can follow the journey in your mind almost without thinking about it. You want to move from station to station in your Memory Palace in the same way you move from your kitchen to your living room. We base our stations on elements that we are familiar with for the precise reason that we don't have to think about what comes next along the journey. We just mentally go there.

And for true success, it's important to think about the appropriateness of the Memory Palaces you choose. As a student, Memory Palaces should always be project specific. This means that you design them in response to specific memory needs. If you need to memorize a number of mathematical formulas, for example, you'll need to put in a few moments of thought. Where do you know that would best serve for memorizing math? What about Shakespeare? History? Scientific facts?

Trust me: after years of doing this, I know that it makes sense to put some thought into what purpose your Memory Palaces are going to serve. Plus, because you draw them, you can be scientific and test which kinds of Memory Palaces work better in general, and which work especially well for certain subjects.

Here's a key point: what you measure improves. Measure your memory efforts by incorporating drawing your Memory Palace floor plans and listing the stations and you'll see your progress multiply, if not explode.

Do This Now

Create your first Memory Palace using a familiar location.

Give it at least ten macro or micro stations, or a combination of the two. Don't fear the adventure if you want to go whole hog!

Make sure that you don't trap yourself or cross your own path as you mentally journey through the Memory Palace and lock those stations down.

Draw the Memory Palace.

Keep a top-down record.

"Rehearse" the Memory Palace journey in your mind in order to ensure that it "works" for the steps to come.

And if at any time you have questions, feel free to contact me at learnandmemorize@zoho.com.

Chapter Three: Bringing Memory Palaces & The Basic Principles Together So You Can Memorization Any Number, Equation Or Formula You'll Ever Encounter In Your Life Again

In the previous chapters we've covered the use of associative-imagery and the construction of Memory Palaces. Now it's time to bring the two together.

In truth, there isn't much to say.

Once you've got well-constructed Memory Palaces under your belt, the only thing to do is:

1. Code number information using associative-imagery
2. Place that imagery on/beside/in/at stations in your Memory Palaces

Of course, there are different kinds of math, so we should talk about these.

The Times Tables

Many of us struggle with multiplication. We're often quite good up until 6 and 7 times 8 or 9 rolls around. Some kids have a hard time getting started with even the simpler configurations.

There's a solution.

Let's imagine that your daughter or son struggles with the times table.

One thing you can do is to help your child understand how Memory Palaces work using your home.

Then, ask your child to explain which numbers present the greatest difficulty.

Moving from station to station, help the child find ideas for memorizing the outcomes of the equations.

For example, let's say that you start your journey in the kitchen. You child needs to remember that 3x3=9.

For kids, simple rhymes can be effective, especially if you make them visual. Here's one:

3 times 3 got a fine, paid the judge, the fine of 9.

This is a sample example only. Encourage everyone with whom you discuss these techniques to come up with their own imagery.

But as we've noted before, it's important to see the images in your mind and locate them in a Memory Palace so that you can revisit them later. You cannot decode the information if you don't know where to find it. This is one of the reasons why having a linear journey is so important.

To give you another example, (even though you do need to come up with your own), take this one:

4 times 4 out buying shoes, why buy one not 32?

Remember when we talked about the Major Method? Here's an example of where you can use it.

Let's say that you have placed this image (either for yourself or as part of helping a child) in the hallway. Since you now know the Major Method, why not come up with an image for 44 and 32. Although you're not dealing with 44, but 4x4, you can still create an image for it and understand that 44 means 4x4. This is something you'll need to experiment to see if it works for you.

Finally, (to obey the rule of threes), imagine that you're now in the living room. Perhaps you or your child sees two melting snowmen fighting over a jar.

8 times 8 (two snowmen) fell on the floor, pick it up, it's 64 (ja + ra = jar).

To some, this might seem like an insane amount of activity just to remember simple multiplication outcomes. And certainly some people won't need any of this.

But for those who struggle, it is imaginative methods like these that can end the sweat and tears and make math fun.

Let's move on to formulas. Assuming that you'll be placing this information in a Memory Palace, imagine that you need to memorize e=mc^2.

You now have more than enough information about using the Magnetic Memory Method to memorize letters and numbers.

But what about that curious symbol? It's called a caret and signifies exponentiation. When I needed to memorize this, I placed it in the bathroom. And there I saw ...

Einstein with garden shears cutting the McDonald's symbol in half. It had one cat ear ^ on it and its other ear was a 2.

What about c2 = a2 + b2?

Before you read on and examine my process, give it a try. What could you dream up, knowing what you now know, to memorize this equation?

For me, the first thing that came to mind was three nuns at the sea looking at the abs on some men walking by. In the grammar of my Magnetic Memory Method vocabulary, the image translated to:

"See three nuns abs."

The nuns, as I'm sure you'll recognize, have to do with the Major Method. For your benefit, let's look at that system once again:

0 = sa
1 = ta or da
2 = na
3 = ma
4 = ra
5 = la
6 = cha, jah or sha
7 = ka
8 = fa or va
9 = pa

This is why the sight of three nuns reminds me that there are three twos in the target equation. The fact that these nuns are at sea reminds me that the equation beings with "c" and the "abs" remind that "a" and "b" are also linked with the number 2.

Don't worry. I can hear the question you're now asking:

What about "=" and "+"?

I suggest that you make a standard image for these that you will always use.

"=", for example, could be vampire fangs. The "+" could be a cross.

Thus, you could image "see three vampire nuns at the sea dragging a cross while staring at abs."

Let's go one better.

Let's say that you also wanted to memorize that this equation represents a Pythagorean triangle.

What do you think you could do?

I'm tempted not to tell you, just so that you'll come up with something on your own. But since we're nearing the end of this chapter, I'll reveal what I would do anyway.

I would see a python in the distance building a pyramid.

But what I see is not at issue.

What matters is what you see.

And what matters is that you create images that are large, bright, vibrant and exploding with crazy action. You want to create a "rubberneck" effect. This will "force" you to look at the imagery when you come across it in your Memory Palaces. It will be impossible not to "decode" what the imagery means, almost effortlessly placing the memorized information in your hands.

No equation or formula is too complex. With practice, you can use the Magnetic Memory Method to master anything you want to memorize.

And to prove it, I have a special gift for you.

Robert Ahdoot is an accomplished mathematician. He's also the man behind Yaymath.org, a popular math-learning resource from which you can benefit at any time. And now even more so with the memory skills you've learned.

But just because Robert's a math whiz doesn't mean that he has formulas memorized.

Quite the contrary. As he told me, he writes out formulas for students from the textbook, not from memory.

Complex formulas.

Twisted formulas.

Long, snarling and daunting formulas.

Guess what?

I helped him learn how to memorize even the most complex of these. In fact, he memorized 9 extremely complex formulas in just 45 minutes using a Memory Palace.

And then we made a video together in which he describes exactly how he did it. And he describes how incredible it felt to make such an achievement.

I'm going to give you that video.

Here's the link. This is from my private Dropbox account so please do not share it. It also includes the transcript of the interview, which you'll also find at the end of this book:

https://www.dropbox.com/sh/o6a3dbanq9gbik6/AAALas FA4iPSxcR2Om3VdEpCa?dl=0

Have you watched the video?

You have?

Good. Inspiring, isn't it?

Here are the most important points to consider about Robert's Memory Palace work with these formulas:

1. He used a familiar location with meaning for him.

2. He drew upon narrative elements from his real life. These elements involved drama and drew upon familial cliches.

3. Robert filled the images along his journey with vibrant and intense action.

4. Robert packed his journey in a compact manner. In some parts, he moved from one chair at the dinner table to the next.
5. Robert reviewed the material using Recall Rehearsal. This process involved a minimal amount of time and ensured that the formulas eased their way into long-term memory.

It is on the matter of Recall Rehearsal to which we now turn. This technique will not only ensure that the numbers and formulas you've memorized go into long-term memory. It will also exercise your imagination so that you can memorize faster, memorize more and improve your mind.

Chapter Four: How To Get Even The Most Difficult Numbers, Equations & Formulas Into Long-term Memory Using The Simplest And Most Elegant Memory Technique In The World

The techniques you've learned thus far make it possible to memorize any number or equation with speed and accuracy. You've also learned how to create a Memory Palace and use it. You, your child or any math student living under your roof can now memorize the times table with speed and accuracy.

But the extent to which the memorized numbers will last depends on a lot of factors. The easiest way to explain these factors is to look at some theories and concepts of memory. Then I will teach you about "Recall Rehearsal" so that you can place any number of formulas into long-term memory. Having done this, you can rest assured that the information will be there when you need it.

Whom is long-term memory for? It is especially needed by students wishing to pass formula-driven exams. It is also helpful for those who perform calculations as an employee. Or perhaps you're a self-employed computer programmer who would enjoy the edge of having formulas on hand. Being able to pluck them from memory saves a lot of time compared to searching Google or rifling through books.

With these benefits in mind, let's see what you can do to get any math information you need into long-term memory.

Hermann Ebbinghaus

Hermann Ebbinghaus (1850-1909) performed many memory experiments. His findings are useful for those of us interested in practicing memory skills at the highest levels. You can find his ideas in a book called Über das Gedächtnis, or Memory: A Contribution to Experimental Psychology.

In this book, Ebbinghaus suggests that learning and retention degrade based on time and position. In other words, the order in which you learn something affects how you will keep it. Thus, the more time you spend on information, or the more "primacy" you give, the greater the chance it will enter long-term memory.

The problem is that we tend to give more primacy to the information we learn first. Ebbinghaus called this the "primacy effect." We get tired, our attention wanes and a whole host of distractions interrupt us. Even the first piece of information we've learned can prove disruptive because it may be so interesting or useful. Our interest in the initial information interrupts our ability to focus on the next piece.

Another term Ebbinghaus uses is the "serial-positioning effect." For our purposes, this term amounts to the same thing, but we'll revisit it again further along because we can "hack" it. The procedure you'll learn will enable you to work memory miracles. Using this special technique, getting mathematical information into long-term memory will be easy.

Why is this important to number memorization?

It's important because we're using Memory Palaces. This means that we're not only learning information in order, but also memorizing it in order. And because this sometimes involves long strings of numbers or formulas, we will suffer from the "forgetting curve." This related principle, also from Ebbinghaus, tells us something important. If you do not practice information you have learned, over time you will forget it ("use it or lose it").

But this doesn't have to be the case. Here's how:

I call this exercise "Magnetic Memory Method Recall Rehearsal."

When you use it, you are literally rehearsing what you've memorized as if it were a stage play.

A lot of people think of the mnemonic associative-imagery as movies, but I think this is incorrect.

Why?

Because movies are the same every time you watch them. Only you change.

But when it comes to moving through a Memory Palace, the images are never quite the same. You are using the combination of location, imagery and action to trigger recall. This lets you "restage" the image-stories you've created. It is a play. And it's also playful when approached in the right spirit.

Quite frankly, in my not-so-humble, but always Magnetic opinion, if this isn't fun, either you're doing it wrong, or mnemonics simply isn't for you. I'm sorry to sound brutal, but usually people haven't gotten the method down and that's why they struggle.

You will eliminate much effort if you've taken care of the following:

* You've created your associative-imagery correctly.

* You've placed it in well-constructed Memory Palaces.

In fact, get these two things right and everything will be elegant, easy, effective and fun. For more help, I recommend that you watch my free, four video series, Memory Palace Mastery. If for any reason you cannot click that link, just type in www.magneticmemorymethod.com.

With all this said, the only thing you have to do when it comes to Recall Rehearsal is to find yourself a quiet place and go through the material. Start at the beginning of your Memory Palace journey and keep going until you've come to the end.

You can do this mentally, but I recommend that you have a pen and pencil. Write everything down from your memory. Take care that you've removed yourself from the source material. Don't have your textbook anywhere in sight so that you won't be tempted to check your accuracy until later. Your goal is to exercise and test your memory.

When finished, only then check your accuracy. If you find any flaws in your recall, use what I call the principle of compounding.

Back to testing your images, this stage of Recall Rehearsal is simple. Once you've written everything out, go back to the associative-imagery you've created. If you've found problems, either add new material, streamline it or make it bigger, brighter and more colorful.

Then Test Yourself Again.

When you're satisfied with your accuracy, use the Rule of Five. This will reinforce the material for long-term memorization. The Rule of Five comes from World Memory Champion Dominic O'Brien. He suggests the following review scheme:

First review: Immediately
Second review: 24 hours later
Third review: One week later
Fourth review: One month later

Fifth review: Three months later

Personally, I think you'll benefit more by reviewing more often than this. Even so, O'Brien's basic layout is valuable and you should keep it in mind.

My Suggestion Is That You Work Like This:

First reviews: Immediately, one hour later, three hours later, five hours later.

Second reviews: The next morning, the next afternoon, the next evening.

Third review: Once a day for each day of the following week.

Fourth review: Once a day for a week the following month.
… and from there on in, keep reviewing at least once a month, if not more often for as long as you want to keep the information intact.

If that sounds like a lot, it isn't. Depending on the amount of material, you can rehearse dozens of formulas within 15 minutes or less. Beginners will need a bit more time, but the speed and accuracy you can build by following O'Brien's or my version of the Rule of Five is fast. Dedicated practice based on an understanding of the principles is all you need.

The reason Recall Rehearsal is so much more powerful than using index cards and rote learning is this:

Instead of using the "blunt force hammer" of repetition out of the void, you are using your imagination. This strengthens not only your memory, but your creativity as well. And the more you do this, the better and faster you get. Not only that, but you learn more. And the more you learn, the more you can learn. This is because you'll have more stored information in your long-term memory with which to make connections.

Finally, to deal with the forgetting curve and to hack the primacy effect, do the following during Recall Rehearsal:

* Travel your Memory Palace journeys forwards
* Travel them backwards
* Travel them from the center to the beginning
* Travel them from the center to the end
* Travel them by leapfrogging forward and backwards

By taking time to do this during your Rule of Five routines, you'll ensure that the information enters your long-term memory fast.

Give these techniques a try and be sure to tell me how you do or let me know if you have any questions by emailing me at <u>learnandmemorize@zoho.com</u>.

Chapter Five: How To Overcome Procrastination So That You're Actually Using The Techniques In This Book And Making Massive Leaps In Your Understanding Of Math And Acing All Of Your Exams

This chapter will be useful for anyone memorizing math of any kind. Without true understanding, even the simple technique of using Memory Palaces can seem drab. Worse, it can feel downright unexciting. If you struggle, this chapter will put you in control of how you approach memorization and Recall Rehearsal.

Then, in the second part of this chapter, we'll talk about more about the principle of compounding. This will help not only your retention of the math you've memorized, but also troubleshoot any recall issues you may be having.

Generating Excitement

I once read Mike Koenigs on speed-reading. For him, one of the best methods for speed reading a book is to pretend that you will be interviewing the author. Not only that, but the interview will be taking place on live television the next day. Millions of viewers will be watching, meaning that you'll need to know the book well. You'll need to have a depth of understanding and accuracy about the specific details of the content.

I think Koenigs' idea is brilliant and adaptable to memorizing math principles and formulas. When learning and memorizing math, for example, you can pretend that you have a book to sell. You know that people are only going to want to own it forever if you are able to win their hearts by speaking to them intelligently. You need to explain the math you've memorized in clear, crisp terms. To amp things up, when I use this technique, I sometimes pretend that a movie deal is in the works. But it will only happen if I can convince the producer that I know math well enough to consult on the screenplay and production.

I know this sounds bonkers, but such "Jedi Mind Tricks" can work well. They create excitement, motivation and urgency.

There are many motivational tricks like this. Anyone can explore them. Once you begin, you'll find tricks that get you excited. Yes, even if you don't naturally feel motivated to learn and memorize math.

Just take these ideas, put them in place and experiment with your own. Track your results and then rinse and repeat what works.

Now onto the job of ...

The Principle Of Compounding

We've already covered this, but it's worth going into more detail to ensure that you've got the full picture.

When memorizing information, you may discover that you cannot perfectly recall one or two items. You feel sure that your images are vibrant, well-located and buzzing with action and energy. Yet, when you look for the math principles or equations, you still struggle to recall them.

This hunt for the material can lead to stress and anxiety. These feelings will make you self-conscious and increase the struggle. You don't want this when taking a test and the thought of stress alone will make you even more self-conscious.

Relax. Refuse to be frustrated or concerned. Any slips in your Memory Palaces are actually opportunities. When approached with the right mindset, they will make you a better memorizer, and you'll be studying the math at the same time you compound, increasing your math knowledge.

When compounding, many of my readers replace the original images they've created. I caution against this because doing so can leave "fossils" that will only confuse matters later.

The more popular term for this "fossil" problem in the mnemonics community is "ghosting." However, I dislike this term because our memorized material should not become ethereal when it dies. If it must fade, it should leave a fossil behind that we can "pour" energy back into.

So when you encounter associative-imagery that needs work, add to the images and actions to enhance them. This will improve your recall rate.

As always, please remember that having action in your associative-imagery is key. It makes the target information more memorable, and the more memorable it is, the more readily available for recall it becomes. The good news is there are many ways to compound images to make them more memorable, especially when you relax while you work.

With that said, please realize that there is nothing wrong with your mind if you find weaknesses in your Memory Palace systems. It's just a matter of going back and compounding the images. In most cases, a second pass will do the trick. Any more than three passes suggests that you need to go back and review the central tenants of the techniques taught in this book. Or you can take my free video course. Just visit www.magneticmemorymethod.com to get started.

Finally, if you want to succeed with memory techniques, avoid rote learning at all costs. The point is to rely solely on your imagination. There aren't going to be any books or index cards around when you're completing a test or examination. During that test it is just you, your imagination and the ways that you've used it to learn and memorize math.

In addition to compounding your associative-imagery, you might like to compound and reinforce the Memory Palaces themselves. This is as easy as popping into the Memory Palace and amplifying it as you would associative-imagery. If your memory of some locations is not as strong as you originally thought, work with another location altogether to form a better Memory Palace. Memory Palaces are in abundance, so if you feel like you're running out just review the earlier parts of this book. I give you many ways to find dozens of them.

Ultimately, the amount of time spent on rehearsing, compounding and "renovating" your Memory Palaces and the associative-imagery you place within them depends on your level of experience and general enthusiasm for memorization. Again, make sure that you complete the preparation and predetermination exercises as fully as possible. Giving them their full attention will save you plenty of time and sweat later. But, when leaks in the system do occur, no stress. Simply wander through your palaces and make "repairs."

Some Common Questions From Readers

Some of the questions that I receive on a regular basis include:

* What do I do if my visual and/or conceptual imagination is lacking?

* What if your representative examples don't work for me?

*What if I don't have images that so conveniently match a mathematical principle?

In the first instance, please go back and reread the chapters on the main principles of the Magnetic Memory Method. These give you several ideas for improving your visual imagination.

To revisit these ideas, a visual imagination is best developed by learning to draw, by looking at art, by building pools of famous actors/artists/sports celebrities/etc., and by actually practicing these methods. Don't overthink the process. Getting started and keeping going will teach you more than anything else. Plus, there are resources like Wikipedia. This alone will give you more than enough art to study. It also features lengthy databases of actors/singers/politicians and all the people you could ever hope to include in a Memory Palace for the purposes of assisting your memorization and recall of mathematical principles and rules.

Second, my representative examples are not designed to work for you. I have given them so that you can model the process. Many would-be Memorizers are unwilling to create their own associative-images and spend hours scouring the Internet for "mnemonic examples," or they try to get the examples they read in memory books like this to work.

This approach confuses activity with accomplishment. Your goal should be to learn how to create your own mnemonic associative-imagery. You then learn to exaggerate the images so that it creates memories that you cannot help but recall, even if you tried.

Some people think I'm a little hardnosed about this stuff. The truth is that Yoda in Star Wars was right.

"Do or do not. There is no try."

Another phrase that has helped me many times over the years is this: "None of us works as hard as we think we do."

Now, you might be thinking: Wait a minute. Throughout this book you've been talking about how easy and fun all this is.

Great observation.

And it remains true.

At the same time, effort is involved.

But that's not a bad thing. People often mistake "effort" for "work." They don't realize that eating chocolate takes effort. Kissing takes effort. Everything takes effort.

Everything has to do with how you approach the game. I suggest that you approach it ...

Magnetically.

Chapter Six: How To Defeat Procrastination Forever

Math students often complain that they cannot focus. Or they claim that they haven't the will power to spend the necessary time on learning.

To address this problem, here are a few points about learning and concentration that I have picked up over the years.

One way of thinking about learning and memorization is to see them as two different skills. By the same token, learning is memorization and all memorization is learning. The only question lurking here is: do you have to understand what you've remembered in order to remember it?

The answer, of course, is no. Many times I have learned a word or formula and forgotten what it meant or how it should be used. As discussed in a previous chapter, this is why compounding images and rehearsal or revisiting the palaces frequently is so important.

Yet, there are barriers that prevent us from taking these important steps. One of the biggest impediments is procrastination. We all procrastinate, and this is just something for the sake of sanity that we have to admit to ourselves. Since we all do it, there is really nothing to be gained from punishing ourselves or feeling bad about our procrastination. The fact of the matter is this: Sitting around feeling bad for doing nothing inevitably leads to more sitting around doing nothing. It makes the problem worse.

The author Tim Ferris, who made his claim to fame with books such as The 4-Hour Workweek and The 4-Hour Body discusses a very interesting method for dealing with procrastination. He allows it to happen. He knows it is inevitable, so he plans for it. One of the best quotes I've heard from him is that we should "budget for human nature instead of trying to conquer it."

The point is that we mustn't punish ourselves for skipping a few days here and there. As Ferris suggests, we will do much better over the long haul if we routinely schedule the days we miss. Intentional procrastination can even be inspirational.

Why?

Because as you are working, you know that some vegetation time on the couch is just waiting for you to enjoy.

For more valuable tips on breaking the procrastination habit, join the Magnetic Memory mailing list by visiting www.magneticmemorymethod.com. A wealth of free material awaits you.

Chapter Seven: Two Relaxation Secrets For Studying Math That Condition You To Excel In Tests And Exams With No Stress, Worry Or Suffering

Harry Lorayne has pointed out that one of the reasons why we can't remember the names of people we meet is because we haven't paid attention to them in the first place. He's right. I believe that tension, stress and not being present gets in the way of the attention needed for Memory Palace work.

The number one reason you want to be relaxed when you learn math is because it will train you to be relaxed when you are trying to recall the principles and formulas in an exam setting. Nothing is worse than knowing something but being unable to recall it due to nervousness or feeling like you are on the spot.

To that end, I want to share with you some principles of breathing that you can use while memorizing math.

We need relaxation in order to overcome such boundaries since so many of us experience confidence issues around our memories. Fortunately, this is easily done.

The two main strategies I use have wider applications than memory work alone. I recommend using them every day for general health as well.

I know of nine breathing techniques overall, one of which I will discuss in this chapter. It is called Pendulum breathing.

The second involves progressive muscle relaxation.

Pendulum Breathing

If you've ever seen a pendulum, then you know that there is an interesting moment at the end of each cycle. This happens when the pendulum seems to hang for an instant. Then it moves a little bit more in the first direction before falling back the other way. It does this back and forth.

Pendulum Breathing works much in this way.

To start with Pendulum Breathing, fill your lungs normally, and then pause slightly. Instead of exhaling, breathe in a little bit more. Let the breath out naturally and pause. Then instead of inhaling after the initial exhale, exhale out a little bit more. By circulating your breath in this way, you are "swinging" the air like a pendulum.

This practice will reduce stress over once you are used to doing it. Make it part of your daily practice while walking or sitting at your desk. If you do nothing else, implement Pendulum Breathing in your memory work. This method of breathing makes Memory Palace construction and the generation of images and associations so much easier.

Why?

Because you are putting yourself in a kind of oxygenated dream state.

At first, it may seem difficult to concentrate on both your breathing and doing imaginative Memory Palace building. This is because in some ways, it is like being a drummer who is creating three or four different patterns, one for each limb. But with practice, the ability will come to you.

Progressive Muscle Relaxation

Progressive Muscle Relaxation is relatively well-known, and yet so few people practice it.

The work is simple: sit on a chair or lie down on a bed or the floor. Next:

1) Point your toes upward and hold.

2) Point your toes towards the wall and hold.

3) Flex your calves.

4) Flex your thighs.

5) Flex your buttocks.

6) Flex your stomach muscles, lower back muscles, chest and shoulders (all core muscles).

7) Flex your hands, forearms and upper arms.

8) Flex your neck, your cheeks and the muscles surrounding your eyes.

Practice Pendulum Breathing as you do this, or at least work to conjoin the flexing movements with your breathing.

Once you have achieved a profound state of relaxation and all of your Memory Palaces have been built, sit with the material you wish to remember. If isolating the principles and formulas helps you, prepare an index card for each.

As ever, avoid rote learning at all costs. Let your Memory Palace skills do the work. Compound your images when testing routines reveal weaknesses. Just as you would relax to remember, relax to test and relax to compound as well.

Again, realize that you want to practice relaxation during memorization so that you condition yourself to be relaxed when accessing the numbers and formulas later during tests and exams. This is the key to Memory Palace Mastery

Conclusion

There is much math to learn and memorize as you continue your adventures with the Magnetic Memory Method.

Because a solid understanding of how Memory Palaces work is critical to your success, let's conclude with some intensive review. We'll also expand on some of the most important Memory Palace concepts as we go through what you've learned.

The first step is to create a journey, but not just any old journey if you're using the Magnetic Memory Method. Instead of simply creating a helter skelter path throughout the building you are using, obey these four principles to create effective Memory Palace journeys that will be fun to use:

* Don't trap yourself

* Don't cross your own path

*Peer versus enter

* Select your "stations" with care

Let's review each of these principles in detail.

1) Don't Trap Yourself.

Over the years, I have found that many people I've worked with wind up trapping themselves in their Memory Palaces. This is because they start anywhere in their home at random instead of thinking the journey through.

For example, I'm presently writing in the kitchen. But in this home, the kitchen would not be an appropriate starting point in this Memory Palace. This is because in order to have more than two or three stops along my journey, I would have to move deeper into the Memory Palace.

On the contrary, we want to move outward, towards an exit. This is so that we can get outside and add new stations or stopping points along the journey at any point we wish.

We always want to be able to add more stations.

Although a subtle point for true Memory Palace aficionados, we also want to avoid "Memory Palace Claustrophobia." This condition describes the feeling that there isn't enough space for the images we have created and left at a particular station.

I would be saying this tongue in cheek, but I have actually heard from one of my readers that this is a problem for her. And I believe it! Anything that causes you to concentrate on matters other than the information you've stored in your Memory Palaces needs to be avoided. Not trapping yourself along any point of the journey is a good place to start.

2) Don't Cross Your Own Path.

This point is strongly related to the point about not trapping yourself. If you have a computer nearby, I discuss this at length in a free video on YouTube I created to help a reader who sent me a map of one of his Memory Palaces:

http://www.youtube.com/watch?v=IQ6j5d7Dvgo

(If you're reading the print edition, or listening to the audio edition of this book and don't feel like typing this address, just search for "Metivier YouTube avoid memory palace confusion" and it should pop up).

On the map, this reader shows how in order to move through his house, he felt he had to cross his own path. However, as you'll see based on the drawing he supplied, we found a solution together.

You can easily follow these two principles by creating your first Memory Palace station in a "terminal location." This term indicates the innermost room in your home that you can move outward from throughout the dwelling towards a door leading outside. Main bedrooms situated in the back corner of a home usually fit this description. In the first home in which I created a Memory Palace, my office was the terminal location.

The path I created was as follows:

My office
Laundry room
Bathroom
Bedroom
Wife's office
Living room
Hallway
Kitchen

As you can see, it was possible for me to mentally move through the Memory Palace in a linear line without crossing my own path or trapping myself. I also did not need to pass through walls like a ghost, nor did I simply jump through the Kitchen window out onto the street.

We avoid movements like this because such actions require mental energy. Unnatural elements create "blips" in your journeys. You will not want to deal with interruptions like these when you use your Memory Palaces later to recall information. Keep the journeys simple, linear and based on what you would do in reality.

I should point out that you don't have to follow my advice here. I'm making this recommendation based on years of experience of my own and countless interactions with readers of my books. They confirm that passing through walls is the equivalent of crossing your own path because it distracts from the primary goal, which is finding information you've memorized.

Yet, it is important to experiment on your own. It is impossible to rule out that such unnatural strategies won't work for some people. I am providing tried and tested guidelines, but ultimately each person needs to adapt the principles to their own use. But if in doubt, move through your Memory Palaces in the same way you would if you were to walk through them along a linear path along which you do not cross your own tracks.

3. Peer versus Enter.

Of course, if you're moving from room to room, how on earth do you avoid not crossing your own path, especially if you want to use multiple places inside of each room to store memorized information?

The problem is easily solved. Instead of entering any room, simply imagine that you are peering into it. If you identify and create multiple micro-stations within the room, instead of walking from station to station, simply cast your eyes (in your imagination) around the room. There should be no need to enter it.

The important point is that you want to make sure that you circle the room clockwise or counter-clockwise depending on the linear progression of the rest of your journey.

4. Select your "stations" with care. Instead of calling each location within a Memory Palace "loci" (Memory Palaces are already locations), I call these stops along the journey

"stations." And these stops literally are stations where you leave the information you've encoded using the other strategies discussed in this book.

There are at least three kinds of stations and a person using the Magnetic Memory Method could certainly identify more. These are:

* Macro-stations

* Micro-stations

* Virtual stations

A macro station is an entire room. If you use your bedroom to store one piece of information, then that is technically a macro-station. However, if you use the dresser, the window sill, the left bedside table, the bed, the right bedside table, the closet and then the bookshelf before exiting the room, then these are all micro-stations within the room and the room itself no longer technically qualifies as a station at all. It's simply part of the route where you pause and peer in the door to take a journey with the eyes in your mind around the room.

Here's A Full Review Of
How To Get Started Building Your First Memory Palace:

1. Identify a location with which you are deeply familiar. At this point, you should use a building to which you currently hold some connection. Again, it doesn't have to be your home. It could be your office or your school. However, avoid things like large campuses. Use a relatively contained structure with a number of rooms connected by hallways and/or staircases.

2. Find 10 "stations" within the location, which is now officially a Memory Palace in your mind. You will use these stations as "drop-off" points for the information you want to memorize. A station can be an entire room or just part of a room.
I recommend starting with entire rooms at the beginning. But if you feel ready to "peer" into rooms by giving them multiple stations, by all means do so. You will learn about your thresholds and limits as you explore the Magnetic Memory Method. And as you explore, your mental abilities will extend.

3. It helps to draw out the floor plan of the Memory Palace on blank paper or graph paper. Again, visit

http://www.youtube.com/watch?v=IQ6j5d7Dvgo

or search for "Metivier YouTube memory palace confusion" and you'll find a video depicting exactly how one of my readers has drawn out his Memory Palace and how to troubleshoot a small problem he had.

As an alternative to drawing out your Memory Palaces, you can also create a top-down Excel file. I usually do both, but it depends on the purpose for which the Memory Palace is intended.

To see an example of how you can use an Excel file to keep record of what you've done in a Memory Palace (including the Memory Palace itself), visit:

www.youtube.com/watch?v=UMPMuOyfke4

(or search Google using the keywords "Metivier YouTube Excel file Memory Palace).

Whether you draw or use an Excel file (or both), number each station in the Memory Palace in sequential order. Ensure that your journey starts in a terminal location (i.e. you've eliminated the possibility of trapping yourself within the Memory Palace). Plus, make sure that your journey moves in a linear line without crossing your own path.

4. Do all of these activities in a state of relaxation. Revisit the chapter in this book on the role of relaxation in imaginative Memory Palace work (i.e. play) if needed.

5. Test your Memory Palaces. Mentally wander through them and make sure that you can move from station to station without spending undue focus on the journey. The

journey should be natural and closely resemble how you would move from station to station if you were really going to walk through the building.

6. Amplify your Memory Palaces. This means that you take a small amount of time to concentrate on your journey to make sure that it is vivid in your mind. A lot of people skip this step, assuming that because they are so familiar with the locations upon which they base their Memory Palaces that this isn't important. In many cases this is true.

However, personal experimentation and the feedback I've received from those experiencing monumental success from the Magnetic Memory Method demonstrates that taking just a few seconds to mentally walk through the Memory Palace and concentrate on the colors, the lighting and even the materials along the way greatly enhances the Magnetic "stickiness" of the Memory Palace. Personal experience will undoubtedly demonstrate that this is true for you too.

One very interesting reader and a participant in my video course, "How to Learn and Memorize the Vocabulary of Any Language," shared the experience that her Memory Palaces were intensely vivified by walking through the Memory Palace and running her hands along the walls. I've experimented with this myself and it works gangbusters. Depending on the layout of your house, you can do this with your eyes closed for extra imaginative benefit.

Once you've gone through this procedure once, you can do it again and again. And because you now understand some of the basic principles behind truly effective Memory Palaces, you can be certain that the information you store in them will be easy to access each and every time you stroll through a Memory Palace in your mind.

More On Gathering Memory Palaces

One of the many elements distinguishing the Magnetic Memory Method from other trainings is my emphasis on creating lots of Memory Palaces and then organizing them in a particular way.

The classical method of organizing multiple Memory Palaces involves a "Grand Central Station" Memory Palace. Imagine, for example, using your high school. In effect, high schools are a collection of rooms connected by corridors.

When used as a central station for your Memory Palaces, instead of mentally walking into individual classrooms, these doors would lead into different houses you've lived in, other schools you've attended, shopping malls, etc.

I know that this option works well for some people, but I've found that it confuses the majority. You have to remember, for example, which door leads to which Memory Palace, and since there are so many doors and so many Memory Palaces, people both new to the game and filled with experience can get confused.

Ultimately, there is little to be gained from this process of linking together Memory Palaces based on real locations using an invented Central Station.

Why?

As you'll recall, a fundamental rule of the Magnetic Memory Method is that we must reduce or eliminate everything that costs mental energy. When it comes to creating Memory Palace journeys and maintaining our networks of Palaces, using an invented gathering place filled with a variety of doors will certainly cause confusion for many people. This problem and its solution can all be summed up in one simple phrase:

The Less You Have To Remember, The More You Can Remember

It's a paradoxical equation, but it's a fundamental premise of mnemonics that is never discussed. The architecture and principles we are building do have a learning curve, but once the Magnetic Memory Method becomes second nature, it is like a very light software code that floats in the background. But plug it up with too many invented things and then you have to essentially rebuild the Central Station every time you visit it.

The Better Method

If we're not going to use a "Grand Central Station" to connect our Memory Palaces, what other options have we? Undoubtedly, there are countless ways, but I have found that using the alphabet as a structural connector works the best.

First, the alphabet is not a building, and yet it is still a structure. It begins at A and proceeds to Z in a regular and predictable manner. If you find yourself at D, it's easy to figure out that C precedes this letter and E follows. If your mind magnetically zooms to Y, then it is not an enormous feat of mental energy to see that X and Z are its closest neighbors.

But due to the nature of how we are going to assign Memory Palaces to different letters, we will never have an issue finding them because each Memory Palace will be alphabetically labeled.

Construction begins, then, by seeking out twenty-six Memory Palaces, each of which begins with a unique letter of the alphabet. For example, when I first created a 26-letter Memory Palace system, I used shopping malls, my high schools, but mostly the homes of friends. I now have multiple Memory Palace systems (akin to alphabet keys on a chain that are themselves alphabetically arranged according to subject) and here is a representative example that you can use to start thinking about and generating a network of your own:

A: Aberdeen Mall
B: Brock High School
C: Clark's house
D: Dawn's house
E: Eric's house

F: Frank's apartment
G: The Garage (concert hall)
H: Heather's house
I: Ian's house
J: Jessica's house
K: Kane's house
L: Liam's house
M: Paramount movie theatre
N: Northern Face store
O: Owen's house
P: Paul's house
Q: Quinn's house
R: Ryan's house
S: Simon's house
T: Trevor's house
U: Uncle Lloyd's house
V: Valleyview High School
W: Walter's house
X: Library
Y: Yolando's house
Z: Zoltan's movie theatre

Let me offer a few notes on the choices here. Not all of these names represent exactly what they suggest. For example, Zoltan didn't own a movie theatre. He was the contracted janitor who hired me to work there from 12-5 a.m. while I was a young university student struggling to pay the bills while I took the only undergraduate course I could afford that year (thanks Zoltan!)

Likewise, "Yolando" is the nickname of a friend whose real name actually starts with an 'E.' You'll also note that

"Paramount movie theatre" is used as the "M" Memory Palace.

Stretching things in this way is to be avoided, but not denied. This is because the mind will naturally bring you ideas, especially when you build your Memory Palace network in a state of relaxation. It's important not to resist unless you feel that the association is too far out of whack and that you'll have to expend energy memorizing it. As mentioned several times already in this book, unnecessary expenditures of mental energy are to be avoided at all costs.

At this point, you may be thinking that the Magnetic Memory Method is a huge investment of mental energy just to get started.

Not so. It will take you between 2-5 hours to get set up and using the full powers of your imagination to hold, maintain and use a system of Memory Palaces.

If you have any doubts about their power, I encourage you to read this article by a woman named Amanda Markham in Australia who used the Magnetic Memory Method to memorize 200 words of Arrernte in just 10 days:

http://anthroyogini.wordpress.com/2013/11/18/learning-an-aboriginal-language-a-quick-dirty-guide-to-learning-grammar/

If you're reading the print edition or listening to the audio edition of this book, you can also simply Google the keywords: "learning an aboriginal language quick and dirty guide."

What I like about Amanda's article is that she includes examples of her Excel files, which allows you to see how someone has used them to achieve a memorization miracle. Naturally, she has followed the key principles we've talked about so far, including not trapping herself within her Memory Palaces and not crossing her own path.

All of what she says applies to memorizing math.

Where To Find Memory Palaces

We've already touched on the use of living spaces and work places for building and developing Memory Palaces. However, I'm often asked for more ideas and my answer to the question boils down to the following:

Memory Palaces are surprisingly easy to discover. Although you may not be a person like myself who has moved from city to city and moved several times within each city while attending multiple schools and working all manner of odd jobs during my younger years, I'll bet that you've lived in more than one house or apartment.

Assuming you have friends and family, you've also visited countless homes of other people. Your personal history is likely also rife with movie theatres, libraries, museums and if you can think in a structured manner about outside locations, there are also parks, forest trails and neighborhood walks at your command.

Wherever possible, it's good to take a walk around locations that you will use as Memory Palaces to amplify your memory. For example, if you can visit an old school, you won't necessarily improve your memory of the structure, but you'll make the location more vivid – and this means that it will be more Magnetic.

Now that you've learned about Memory Palaces, the next major step is to always keep one simple fact in mind: every place you visit can potentially become a new Memory Palace. You can deliberately focus on the location by paying attention to it in a completely new way, an intentional way that will make the layout even more memorable.

If revisiting locations isn't possible, you can look at old photographs, or in some cases, use Google Earth or Google Maps. In the case of public places, you can often search "blueprints" or "floor plans" and see representations of locations ranging from public libraries to shopping malls to casinos. In fact, I was given this idea by someone who wanted to use a casino he'd once visited and searched the Internet for a floor plan to help reconstitute his memory of the layout.

There are endless ways to revisit locations, and again, keep in mind that if your past happens to be limited, you can always strike out into the future by visiting new locations with a prospector's eye. There is truly no end to the Memory Palaces you can build.

Once you've compiled a list of candidate locations, I recommend filling out the Magnetic Memory Worksheets.

These can be downloaded here:

http://www.magneticmemorymethod.com/free-magnetic-memory-worksheets/

It should take you only an hour or two to complete them. When you've done so, you'll have a 26-Memory Palace network with ten stations in each Memory Palace. Because you are following the principles of not trapping yourself and not crossing your own path in these Memory Palaces, you'll be able to add new stations to individual Memory Palaces later. If you're not using the special, Telesynoptic Memory Palace technique taught in other books I've written (this technique is actually more appropriate to memorizing poetry so please forget I mentioned them unless you're truly interested in the next level in Memory Palace technology), you can also assign more than one Memory Palace to each letter of the alphabet.

For example, you could have:
A1
A2
A3
B1

B2
B3
B4

This strategy can be especially handy when using Memory Palaces to acquire the massive amounts of mathematical principles and formulas.

The Magnetic Memory Method Is Perfectly Suited For That!

In sum, the building and development of Memory Palaces takes only a small amount of time and effort. The next step is learning how to fill the Memory Palaces you've prepared with the information you want to memorize. This could be anything, ranging from facts, lists of historical figures, foreign language grammar or names and faces.

As a final suggestion, as you are filling out the Magnetic Memory Worksheets, concentrate on the journey and make it as vivid as possible. You can literally close your eyes and pretend that you are "turning up the volume" on the Memory Palace.

You can try this in the room you are currently in, reconstructing it in your mind and then making the layout bright, vivid and pumping with energy. It should almost be as if you're casting some kind of spell or attempting to manipulate reality like Neo in The Matrix. And manipulating reality you are.

Next time you are out for a walk, shopping or just wandering around the house, consider the hundreds of locations you can use to build and extend Memory Palaces. The more we pay attention to our surroundings, the more material we have to work with.

As well, take every opportunity to visit places you've previously lived or gone to school. Revitalizing your familiarity with the locations you use to build your Memory Palaces is not entirely necessary, but at the very least, you should perform a mental walkthrough to ensure that you have enough material for at least the first 10 stations and ideally many more.

In addition, utilize the power of your imaginations and the images it brings you. Harness the power of coincidences such as those I related in the examples given in this book.

Make sure to remember the bicycle metaphor for memory and suit the principles to your own needs by making adjustments to the system taught in this book. You should never be afraid to play around, amplify and use absurdities.

Test yourself and compound regularly or when necessary.

And always, always relax when doing memory work.

You should also spend time thinking about the kinds of math principles you would like to learn or need to know.

You should analyze how you can group different rules together and develop your understanding of math based on your areas of interest and goals. You will see many more connections by doing this.

It goes without saying that you should recite the math principles and equations you've memorized as often as you can. Practice Memory Palace recall while speaking with friends or study partners. This means searching for the rules using a specific principle or formula (mentally walking through your Memory Palaces), rather than casting a hook and hoping a math rule swims by and bites.

Finally, teach others what you have learned about memorization skills. Talk about how you built your Memory Palaces, the techniques of location, imagery and activity. Give your friends and colleagues examples of how you've memorized specific lines. Teaching others is one of the best ways to compound information that we've learned and it allows us to see other possibilities and new techniques we may have missed.

From this point on, you are now more than equipped to succeed with the Magnetic Memory Method. I hope that the examples and instruction throughout this book have helped you see the possibilities and options you have for creating images along dedicated Memory Palace journeys that enable you to memorize math concepts. If you have any questions, you can contact us through me at any time. My email is learnandmemorize@zoho.com and I endeavor to answer all questions normally within 24-72 hours.

About the Author

Anthony Metivier completed his BA and MA in English Literature at York University in Toronto, Canada. He earned a second MA in Media and Communications from The European Graduate School in Switzerland while completing a PhD in Humanities, also from York. As the author of scholarly articles, fiction and poetry, he has taught Film Studies in Canada, the United States and Germany. He plays the electric bass and is the author of the novel *Lucas Parks and the Download of Doom* and *The Ultimate Language Learning Secret*.

Be sure to visit http://www.magneticmemorymethod.com for access to the free Magnetic Memory Method Podcast where you'll hear interviews with memory experts like Jim Samuels and Harry Lorayne (subscribers only) and language learning giants like Luca Lampariello, David Mansaray and Sam Gendreau. You'll also find Anthony Metivier's amazing "Memory Training Consumer Awareness Guide," "Memory Improvement Master Plan" and much, much more!

How To Memorize 9 Complex Formulas in 45 Minutes:

Bonus Interview with Math Expert Robert Ahdoot

If you haven't already, as a reader of this book, you are entitled to view this video at no charge and with no strings attached. Here's the link:

https://www.dropbox.com/sh/o6a3dbanq9gbik6/AAA LasFA4iPSxcR2Om3VdEpCa?dl=0

Here's the full transcript, edited for readability:

Anthony: Why don't you just tell us a little bit about yourself and then tell us what you've been working on the last couple of days.

Robert: My name is Robert Ahdoot, the founder of YayMath.org. And I found my way to you through the un-con school. What I do at Yaymath.org is record my math video lessons live in the classroom. I dress up in costumes and Yay Math has been in existence for five years. We have over 5 million views and about 1.5 million minutes viewed per month because people really enjoy the live student interaction with teacher, the spontaneity, the authenticity, the imperfection. And it is through this, that I found my way to you, over the last few days after our initial conversation. Since then, I've learned about what you do and have seen how your practice of helping people use their memory is accomplished through the Magnetic Memory Method.

So I've been learning about the Magnetic Memory Method and I've been trying to put your tactics to use to help me in my practice of memorizing math formulas because I mean, even though I'm a math teacher, there's a slew of formulas I still need to reference and look at my own crib sheets to recall. However, at your suggestion, we're going to be creating a mental crib sheet for me to memorize formulas and that's what I've been doing. Pretty recently, it actually didn't take me that long and I can't wait to share what I've done with you and see what you think.

Anthony: Okay. So say a little bit more about these formulas. What kinds of formulas are they? What characterizes them and what region of math do they belong to, so to speak?

Robert: So they're statistics formulas. Statistics has been my latest craze, my latest passion. It's a sub-section of mathematics that I've been filming most recently and that I'm admittedly the newest to, which has been very exciting to learn on the fly. So I'm using your methods to help me remember these formulas for myself. And not only that, I'll be able to usher this methodology to class after class after class that I end up teaching and give them these same tools so that they won't always be always fretting about what the formulas are or how to use them.

One of the number one questions I get is, "What are all the formulas we need for the test?" They say that. I end up writing them on the board and then just by repetition, I end up remembering them – not always. And sometimes I don't remember them and I say like, I would have to commit them to short-term memory and that's what students do, but using this Magnetic Memory Method, hopefully it'll be committed to long-term memory. And I hope to demonstrate that to you today and it's going to be fun! Can I tell you about the process about what it was like for me?

Anthony: Yeah. Absolutely. But one thing I think would be very interesting for people is, in terms of getting these formulas into long-term memory, what are one or two or even more of specific benefits beyond just passing a test that you could think of that someone is going to benefit from in having this ability?

Robert: Okay. That's a good question. Because I definitely refuse to teach towards tests. I believe that we need some form of assessment for students in general, but I think the confidence that students, such as me in this case, can pick up a process on the fly and completely learn something that they previously had not known nor were able to do well. I mean it's one thing to just write down formula after formula after formula and just try to commit it that way, but the process was very invigorating, I've got to say because I was able to take these things – and it was their story and images. That's some of the things that you talk about. And it was fun! So, not only does it increase the confidence in my own capacity to learn, but it makes the process of learning fun and that's pretty hard to do generally when it comes to memorizing formulas. They're just a bunch of symbols to the person looking at them for the first time. But, turning that sort of process into a game or a stroll down memory lane, it was kind of cool to be able to do that. So I appreciate that.

Anthony: Okay. So, tell us then, about your process.

Robert: Okay. Yeah, the process. I don't know if you've known this and I've been curious to talk to you about this. If you're going into what you call a Memory Palace and you're conjuring up images from your life that are personal to you, in fact maybe very near and dear to you, I think it's just a matter of time before you trip up over some sort of emotionality or even vulnerability. It becomes almost like my process what I'm going to reveal to you today is stuff that I grew up with as a kid. Normally, if you and I were talking – as much as I enjoy your company and enjoy hearing what you have to say – I wouldn't really talk to you about what I did when people got snappy in my house. You know? And that happens in this story.

People get snappy and I'm replying to them in formula. I'm replying to them in formula speak and so there's a level of vulnerability that is required – at least for myself to face this type of stuff and then, furthermore to be able to project that out and share with other people. Have you noticed that, that's normal? It's not necessarily going into a building and floor 1, 2, 3 with an elevator and that makes sense. It could be like something occurred in that building or people were in that building that conjured up something from your past. Is that normal or is that just me?

Anthony: Yeah. It's certainly something that – I haven't really myself experienced that. I'm a pretty neutral guy and sort of scientific about things and for some reason I just have a kind of clinical ability to use the Memory Palace without really running into ghosts so to speak. There are people who have told me that they cannot use certain places because of a history of violence or some sort of thing that has happened in that house or association. And that would be that they're bumping into either a direct or indirect memory.

The other thing that sort of goes along with this is you mentioned that there were things you wouldn't really want to relate that have gone through your mind in order to memorize some of this material. And this is one of the most controversial things about mnemonics and memory techniques and memory tricks or whatever you want to call them is that they do involve certain extreme images that can incorporate sexuality, violence, usually cartoon violence, but violence nonetheless because you're trying to evoke what I would call the rubberneck effect, like a car accident, you simply have to turn your head and look.

There's just something about our ability to imagine things that can be quite shocking. And that, in itself, creates emotional reactions. And that's certainly something that was a problem for me because it's not always the most pleasant thing to think about these sorts of things. But at the end of the day, it's the difference between being able to memorize something and not being able to memorize it. And the fact of the matter is that the mnemonics, the associative imagery are tools and if you do the exercise correctly, they're really in fact, short term and you don't need to obsess upon them or anything.

So the most important thing that I think, and you don't have to go see Dr. Freud to do it, is just simply not to judge your images. Just go with what comes to your mind. Have a clinical distance to them. Don't get too involved, but if there is emotion there and it's comfortable, then definitely exploit it. Leverage it. Use it for all it's worth. And just to quickly add something, the history of mnemonics was deeply suppressed in certain historical periods because of this character. It was considered blasphemy, for example, to use associative imagery to memorize biblical verses because of the kinds of images that are used. So they are certainly things that are very controversial in that area, but my advice to anybody is just to develop a bit of scientific distance from it and when emotions comes that you're uncomfortable with, then use them for all they're worth because what is memorable can be linked to what isn't memorable in order to make it memorable. And that's a very, very powerful tool.

Robert: Nice. Nice. And definitely, I think fortunately I've walked the border or a fine line of exactly what you're saying. It's basically what I would never have really had a need to share with you, but if you and I were sitting over a couple of beers, I would have no problem talking about what happened in my old house. And I plan to right now! So, it's kind of a fun way to get to know me.

Anthony: We'll have beer later.

Robert: Yeah. Or we'll just do it over Skype and just talk about my childhood through statistics formulas. I mean, you've got to say, it's kind of – it's just kind of nifty. So yeah, it has definitely brought up stuff, but nothing in a way that was detrimental. And I definitely leveraged the emotion to get into it. So I have like hand motions. So I'm psyched. Can I show you what I've learned?

Anthony: Yeah. Please, by all means. But let's have a look at the formulas first:

Binomial Probability Distribution

$$P(x) = \frac{n!}{(n-x)!x!} \, p^x q^{n-x}$$

Poisson Probability Distribution

$$P(x) = \frac{\mu^x \cdot e^{-\mu}}{x!}$$

Confidence Intervals:

$$\hat{p} \pm E \qquad E = z\sqrt{\frac{pq}{n}}$$

$$\bar{x} \pm E \qquad E = z\frac{\sigma}{\sqrt{n}}$$

or

$$E = z\frac{s}{\sqrt{n}}$$

$$\sqrt{\frac{(n-1)s^2}{\chi_R^2}} < \sigma < \sqrt{\frac{(n-1)s^2}{\chi_L^2}}$$

Sample size

$$n = \frac{z^2 p q}{E^2} \quad (\text{proportion})$$

$$n = \left(\frac{z \cdot \sigma}{E}\right)^2 \quad \text{mean}$$

Test Statistics

$$z = \frac{\hat{p} - p}{\sqrt{\frac{pq}{n}}} \quad \text{proportion}$$

$$z = \frac{\bar{x} - \mu}{\sigma/\sqrt{n}} \quad \text{mean}$$

or

$$t = \frac{\bar{x} - \mu}{s/\sqrt{n}}$$

$$\chi^2 = \frac{(n-1)s^2}{\sigma^2}$$

Robert: All right. Great. So we moved out of my childhood house many years ago, maybe 10 years ago. I was growing up in Maryland. I figured that was the first place to start if I was going to do this Memory Palace technique to go back to the place where I grew up because I haven't been there in so long and this would be a great way to experience it. So here I come. I figure I'm walking through the door. So stop me if there's anything I can do better. I'm new to this, but I'm trying it out hard.

So I walk through the door and I get to the dining table. That's the first thing there. Something that's going on usually when we're dining is that for some odd reason, we never really had an emphasis on napkins when we were dining. It's like napkins were, oh, yeah, I can use a napkin, but I never really thought to use one until one was offered. So that has gotten me in trouble over my life with manners and things like that at tables as an adult.

So I walk in to this dining table full of everyone, family, extended family and I say to them Napkin! And then, they look back at me and they're like, Napkin X and that's the X I have to think of. They're just saying to me. So they say, Napkin, No. And then they say No for emphasis. And I say back to them, please, please stop with the X. And an uncle that I know, is sort of like a peacemaker and he goes, Quit saying X to the napkin request. So it's like I saw, I want a napkin. They said No to the napkin. They said No twice. I said Please, stop saying X (no) and then someone said, quit saying No to the napkin request, which is sort of played out in my mind there. What do you think? Am I on the right track here?

Anthony: Well a lot depends on what formula that allows you to write out on a piece of paper.

Robert: That was the binomial distribution formula, binomial probability.

Anthony: Okay. And…

Robert: Do you have it with you? Can you follow along?

Anthony: Yes. So the binomial distribution. This Memory Palace journey enabled you to recall the binomial probability distribution – the name of the formula or actually the formula itself?

Robert: The formula itself.

Anthony: So what about the napkin makes – I see there's an N in the formula.

Robert: That's the N for napkin. That's where my mind went.

Anthony: But what does it really represent?

Robert: It represents the number of trials within – it's like trials meaning like flipping of a coin or rolling of a die or something like that. It represents the number of trials.

Anthony: Oh, okay. And then you mentioned the X coming two times. And I see that there.

Robert: Yeah. They say No to the napkin request and they're saying with emphasis. That's why that factorial symbol is like sort of exclamation. It's an easy segue. It's sort of like No! No napkin! No twice.

Anthony: Okay.

Robert: I feel weird.

Anthony: It is weird, but the fact of the matter is that it's working. And one of the strange things, one of the really weird aspects of all of this is that we can explain to other people how that we came to be able to recall something by decoding this imagery. But ultimately these examples are useful only to us. So if you put mnemonic examples into Google, you will find thousands, if not millions, of people who have shared examples. And I think they're a barrier to entry for a lot of people because, okay, napkins, your family, this sort of thing. It only makes sense to you. But people really have to make their own, like you've done. And it's wonderful that you're able to recall that and I assume that, that's going to have a particular function in an exam or in a practical setting where you're trying to calculate something that will enable you to accomplish a goal.

Robert: Right. You're creating the ability for people to do it themselves. That's what you're trying to do. You're not trying to learn it for them. You're trying to help them learn it for themselves. And I get that. I get that. All right. Continue our next question.
Anthony: How does that feel to be able to do that?

Robert: I'm saying that I am an academic so I jumped in with two feet and I will admit that the process was weird because it was against the methodologies that you've already learned before. We're indoctrinated with these learning models, just speaking from one educator to another. We're indoctrinated with learning models that pretty much stop more or less at, are you a visual learner? Are you an auditory learner? Are you a kinesthetic learner? Do you learn by doing or some sort of combination of those? And then there are students who say, I need to be shown or I need to look at it or I need to listen. And we stop there.

What's cool is that this is creating an entirely different genre where you're going into your own mind and taking a stroll and that, to me, is weird, but in a very good way because as long as you have a sense of adventure and you're open for a challenge in something that is new, then I think it can work well. So that was the process. It was like, I was for it, but I had to get over the act of going like (watch video to see Robert's gestures) I was doing this to myself in the room, empty room chilling here. I was like, X X, what am I doing? I think of family feud, the show, that's what comes to my mind. Mnemonics that are only personal. Family feud has that thing, it was like, let's see Eggs and it was like X No. Eggs is not an answer.

Anthony: It raises an interesting point, actually, because there is benefit from actually incorporating physical movement into it. And also actually moving. If you are using a Memory Palace that you actually have access to, there's also a benefit to being in it and moving from room to room. So one of my greatest experiments with that was memorizing the lyrics to a song in German and actually physically moving from station to station in the memory palace as I recited it.

And to what extent that is actually necessary or helped the process, I don't particularly have any hard data or anything like that, but it's just sort of a memorable thing to do and it just adds, it compounds, it gives more oomph to the process, if you can do that. I also heard something from someone and I tried it myself. It was very interesting. She walks through the Memory Palaces that she can actually visit with her eyes closed, running her hands across the wall to increase this sort of sensory, spatial material memory of the place itself which is part of this whole idea of the Magnetic Memory Method, that you're magnetizing things by how you're treating them or using them.

Robert: Right. That makes sense.

Anthony: So let's hear another one.

Robert: Okay. Sure. So my reply to all this X, to my napkin is when I say to them, when you go like X this to me, it makes me have negative feelings about you. It makes everyone have negative feelings about you. So you going like this X makes everyone negative to you. So take it back and back to you. It was like a sort of retaliation. So if you have the formula, you'll see that it's Mu, which is the mean. It's like a U. It looks like a U.

When U go like this X, it makes everyone E, the exponential, feel negative about U and so I retaliate by doing it back to them. X

Anthony: Now I'm seeing very clearly exactly how that's working.

Robert: Then, the one uncle, the peacemaker, he goes, easy, easy, easy. All right? Actually, it's easy, easy, ET. There were three formulas there. The first one there is, EZ, all this please and quit it is over now. Over now. Hopefully you're seeing that.

And the second one, he's saying, EZ again. Please take out your bowl. And then I think I wrote the formula wrong to you. I rechecked. The third one isn't easy, it's ET. So it's ET, take out your soup, or your spoon now. So there was a bowl and a spoon. So he's basically saying it's time to eat. So saying it again, the first easy is all the please and quit it is over now. The second easy is, take out your bowl. The third one is ET, take out your spoon and that's all over now. The second too I know are square root of now. I see it though, that it's over now. Are you...kudos to you that you're able to follow along this crazy narrative.

Anthony: I follow it exactly even though I don't know these symbols and I don't know exactly what that shape of the bowl represents.

Robert: Sigma. Standard deviation, but I needed a symbol like a bowl. You did something with garden shears last time we spoke. And I took a queue from that. I was like, okay, just some kind of symbol while we're eating and we were at a table. So just it makes sense that a bowl and a spoon is there.

Anthony: Yeah. Well that's exactly the sort of thing that you want to be doing, which is substituting, associating and really applying these techniques. And the other thing that's really great that you're doing is you're making a context for it. It's not like a bowl and then Sandra Bullock, but it's a bowl and a spoon.

Robert: Right. Right. Right.

Anthony: And although on the paper I'm looking at, although you have EZ, you mentioned that it's actually ET. And in my mind the first thing I'm thinking of is the dinner scene in the movie ET. Right? I don't recall exactly that there's a bowl and a spoon, but there's something where they're in the dining room with him or even a restaurant I think, at one point. So even my mind as I'm listening begins to work on this and who knows if I'll ever see this calculation again or this formula again, but I may have some recall of something.

Robert: Sure. Exactly. You know that formula and what it – it can be easily translated to you for you to create your own process and I get that. I get that. So it gets really good now for the last three formulas that I memorized, okay?

Anthony: But just before we move ahead, can I ask a question?

Robert: Yeah. Sure.

Anthony: Confidence intervals, you have at the top. Is that just something you don't need to memorize because you know it through familiarity?

Robert: I do. I did. I had that as part of the story. The part of the story was I was trying to work on the narrative, but I stumbled with the act of my gaining my confidence though because it's – you have to know the vocabulary of it. And so at this point I would say creating a different Memory Palace for the vocabulary. Like what is a binomial distribution? What is a poison [ph] distribution? What is a confidence interval? Those types of things. I think you have to know what those are in addition to this, or in tandem with this or I think it's possible if you get good at this to weave confidence interval into this current Memory Palace as well. I'm saying the definition of what it is. You know. It's a lot because you have to know the definition and then you'll be able to do the calculations with the definition.

So I think knowing the definition is probably more important, but it's easier to remember the definition than these formulas. I'm saying it both as someone who teaches it and someone that is trying to learn these formulas. Formulas are a beast. They're so cryptic whereas if you could explain, like you said in our last conversation, being able to explain something that's very challenging in a single sentence. And from the sentence you create an image and then from the image you create your Memory Palace. So it's a different beast to memorize the definition. What do you think about that?

Anthony: Well I think that with greater experience, you may be able to incorporate that as you continue developing. However, that said, I think there's a relationship here between vocabulary and grammar when using Memory Palaces for learning a language. So for example, there are a lot of vocabulary rules that apply to words that look differently in different situations. So one of the questions I always get is do you memorize all of the different permutations of a word with itself or what do you do? And I suggest a separate Memory Palace for the rules in many cases. I mean, grammar is incredibly complex and it's not really something that I've done a whole lot of work on yet.

Nonetheless, there is a benefit to having Memory Palaces or a cheat sheet for grammar rules in a Memory Palace and being able to cross-index them, so to speak. So you have specific instances, like let's say that a word is a formula that has a definition and you can memorize sort of the sound and the meaning of the word at the same time, but it needs to be cross-indexed or so with a grammatical rule in some cases. So if you're able to sort of jump from one Memory Palace to the other, it's almost like Tesla rays or something like that.

Robert: Everyone's fantasy to teletransport. I've always wanted to do that.

Anthony: But in terms of gender for certain words, you can just incorporate certain symbols. One of the examples that I give that I've used a lot is a boxer or boxing gloves is always somehow incorporated into an image with a masculine gender, or a skirt with a feminine gender or fire as part of neutral. But definitely, I think that, that is a very interesting issue that people need to explore on their own and come up with solutions once they know the method. But it sounds as though the real beast, as I think you put it, is the formulas themselves. And you've cracked the code as we've seen from the two of these things. So really the English definition is probably easy peasy next to this.

Robert: It would be easier. It would be easier next to this because it's a conceptual thing. You can explain, and I've explained that numerous times to students that confidence interval is like the interval at which you believe the true population lies. It's like if you do a survey and you say that 80% of people believe in climate change in my survey and then you do some interval somewhere between 80% give or take 2% is the true population percentage that believes in climate change. It's like – that's – once you understand the definition, then the beast is this thing. And I would need to cross-index. I believe in that. I would need to cross-index. Okay.

Anthony: Are these all the confidence intervals or just the ones you selected for this particular exercise?

Robert: These are pretty much – there are three general confidence intervals. One for P, which is a percentage. One for X bar, you see that is for means, for average. And one for standard deviation, the bowl, the sigma. There are three of them there. Basically you do a sample study and then you ask yourself to what degree does my sample apply to the population at large. That's basically what a confidence interval is. And these are pretty much those three, at least with the introductory statistic studies, those three.

Anthony: Okay. Do you want to do some more?

Robert: Sure. I could do some more. I'll do these last three ones and then I wanted to share with you a little brief story about how, unbeknownst to me, I was doing your methods without even realizing it. And before we had even met, I wanted to tell you. The kids love it. It's one of the best videos that people like. So I'll end with that one. But basically, after all this stuff [you've already heard], I'm not really hungry – all this yelling back and forth. Take out your bowl. Take out your spoon. It sounded very sort of like dictating to me. So when I was a kid to blow off steam, I would play Nintendo. And one of my games on Nintendo was Ninja Guiden and it was a Ninja game.

Anthony: I remember that one.

Robert: You remember Ninja Guiden? And so, I just remember because what prompted me was. It's called ki square. It looks like a throwing star. So that's just how I –

that was my in, the throwing star. So it goes like this. Now, one sword. So it's like the minus sign here. Here's the minus sign and the sword, you know, is like this. Now, one sword. And the opponent says, two swords. And then there are two throwing stars. So now one sword, two swords, "s squared" and that's over two throwing stars. And that's twice around the bowl. I don't know why it's around the bowl. It's probably because I didn't want the bowl and I wanted to go play Nintendo. I just made that one up. I just did that one on the fly, but it works.

After I play the game, I'm hungry again. I'm hungry again, so I go back to the kitchen and they present the food. Now, here. See, that's N equals. Now here is the equals. All right. So here we go, more of my past. My parents were born in Iran so I grew up with a whole array of Iranian cuisine. And I'm going to tell you what one of those dishes was, it was call Zettesh Polo [ph]. It's sort of a raisin rice. Okay. So I have zettesh polo. Two servings of the zettesh. That's Z squared. So zettesh polo with quince. That was another thing it was served with. And that's poured all over two eggplants.

Are you following me with the formula where it says it [refers to formulas written on the form]? And then the second dish was one serving of zettesh and then the bowl is back, so you pour it in the bowl and that's over one serving of eggplant. But I'm hungry, so I want that twice. That's the squaring. So again, now, picturing I took a flight over to Germany and I was hanging out with you and we were having the beers. I would have no problem explaining to you that I grew up in an Iranian household with Iranian cuisine and Zettesh Polo was one of them. Zettesh Polo is raisins with rice and quince and that eggplant would've been on the table too. I would have no problem explaining that to you, but had it been for this opportunity I never would've gone there, probably. I would've talked about relevant stuff. So that's what I'm saying.

Anthony: Do you have any alternative ideas that you could use to also memorize this stuff that comes to mind?

Robert: Alternative ideas? How do you mean?

Anthony: You've gone to this particular dish and how it was involved in your culture and so forth. But, if you were pressed to come up with different set of information, do you think that you could have an alternative mnemonic associative imagery?

Robert: Sure. I mean isn't the list infinite?

Anthony: It could well be. But I'm just curious because as we're saying, everyone has their own take, but a lot of people will wonder, what if I don't have anything? So I'm just trying to think…

Robert: So what if I don't have anything? Well I mean, you have to – okay, look, as someone that's new to this, I think my, it was finding where I reside within the palace. What are the images associated? What are the actions? That to me was the challenge. And I would be sort of lost in thought thinking about that. The second I come up with an applicable narrative, an applicable Memory Palace, then I would be able to do it. And I'm saying I'm not – it didn't take that long, but it was definitely a lot of effort. I was sitting here this morning. I was thinking, where am I? What am I doing? Who's around? What are they saying? What do I smell? How do I feel? What's the temperature in the room? These are the things that – and so it's not an easy process, by any means. But, once you're there, it starts to hit. It's sort of like it comes in waves.

But, I wouldn't, without wasting your time, be able to come up with another palace on the spot or something that would make sense to me, like I could go to [audio gap] quail. ZPQ would be zebras and pigs and quail, but they don't really have any particular personal meaning to me, but I would have no problem. And then elephants on the bottom. I wouldn't have a problem remembering it, you know like you said, you call it the training wheels. You use them to usher in the memory into your head and then once it's in there, then you can take off the training wheels. You don't necessarily need to remember zebras and pigs and quails and elephants, but maybe the fact that I've said it three times will make me take off the training wheels.

Anthony: Right. Right. Right. Well, just for the benefit of people watching this, I think that one of the things to be said is, if you're not able to have Z squared with P and Q, associated with something so convenient as a dish, which I actually want to say something about that in a second, but if you didn't have that, you could for instance you gave the example of a zebra, a pig and a quail. Well, you could have the zebra swinging some sort of an appropriate weapon at the pig who is then being attacked by the quail at the same time, or somehow get all these images in there. But the point is, without a personal association, the technique is to exaggerate the violence or exaggerate the action or to make everything big and vibrant and colorful and just zooming with action so that you're creating this rubberneck effect if you're not able to bring anything personal to it. So I think that is a clue for people who are just like, well I don't have any exotic dish that sounds just like that.

But what I do want to say about that is that one of the most amazing things that I've noticed both in myself and others when they start to get into this stuff, I don't have any data on the unconscious mind for this, but it seems as if the unconscious mind seems to arrange things conveniently. And that there seems to be just an absolute overflow of coincidental opportunity to link things.

Again, I don't know how to account for that, or test it scientifically, but it's my impression. It's my impression on a personal level, anecdotal level and an anecdotal level from other people. And it almost feels as if there's sort of a reverse undertow or something like that where the mind prepares things in advance and is just ready there for you as if it knows. Again, I don't want to get into woo woo or pseudoscience or bizarre things, but that's the feeling that I have. And it's just almost too convenient at some points. And I don't know if that's something that people can cultivate or not, but it is something to think about and focus on allowing to happen.

And that's one of the sort of things that you describe. Well, what am I going to do? Where am I? What's the temperature? Who's here? That's a process of allowing something to happen. And when you kind of just get out of the way, it starts to come and you get this kind of effect that I'm talking about, convenience, convenient imagery just sort of popping up. So I'm curious, did you do anything to relax yourself, which is one of the key things I teach in order to enable this kind of flooding up of imagery?

Robert: I did. Again, as a teacher myself, I believe in creating space to learn. And so, I cleared the morning. I had a good breakfast. I slept well. And so once I sat down, I was fully focused and without other sort of stresses of the mind or in my life in the way. I do that as a learning practice automatically. So I'm glad that you make that goal explicit. That's a very, very good explicit goal that you have.

Can I just comment about your claim about the subconscious mind, the unconscious mind? I think it's so interesting and a very interesting claim and makes a lot of sense and my reason as to why, my guess is that we're always in survival mode and it seems like the brain's role in survival is to create predictability, to create order. So it would make sense that it would do that subconsciously in the face of all of this chaos, that it's just ready to go to put things into buckets so that we could have some level of predictability and predictability leads to survival. So I think that that's a very, very plausible claim and one that I'm going to think about a long time after our conversation.

Anthony: There's certainly a lot of related research into the unconscious that they've done through certain tests and so forth that relate to this that definitely can be investigated, something to get into at a certain point. One of the other things that I wanted to ask you about is that, it sounds as if you've managed to do all of this, which is actually pretty substantial, in one room. Is that the case?

Robert: Yeah. I did it in one room.

Anthony: And do you feel that you could go to, say, the confidence intervals information without having to first visit the binomial probability distribution information? Or do you need to start at the beginning?

Robert: Now maybe I could start at any room, but at the beginning, I needed the full tour. I needed to go into the house. I needed to see the dining table first. I needed to go into the family room and play Nintendo and then I needed to go into the kitchen for the dishes. So I think early on, I needed to go from room to room and now, especially with this coaching that you're giving me, I can start in one room.

Anthony: Now, what is your, assuming all of this material – you were able to retain it after spending really what sounds like a pretty insubstantial amount of time this morning…

Robert: Yeah, this morning.

Anthony: … What I want to ask, even though you're not actually going to do this, if you had to produce this information on a test, say next week, and assuming that you maintained your recall through rehearsing it several times between now and then to make sure that it sticks and to re-amplify it, what is that test situation going to look like? What do you expect to see on the exam? And how do you predict that you would use these constructions that you've built inside of a memory palace in order to assist you in passing that exam and being successful?

Robert: Right. That's a good question because many times I think the important part of an exam is not only to know the formula but to know when it applies, on which problem to use it and in which problem to apply it. It's sort of like knowing when to use a screwdriver or a wrench or a hammer. It's the same idea. You have to know when to use which tool.

And so what I would do, is that just given like if I'm trying to embody what it's like for a student to take a test with the pressure and all that stuff and the fear of not remembering I would suggest coming into the exam and just going through my memory palace, going through that process and writing down all of the formulas at the top of the page, just getting all of them down. Getting all of them down would be really important. At least that takes the pressure off from having to recall it at any specific moment.

From there, knowing when to apply which formula to which problem is exactly what you're saying about the cross-indexing because you would need to look for key words within the problems. The problems would say, for example, what is the confidence interval estimate if my sample size if 40 people and my margin of error is 2%?

You know they would give you that N. They would give you the E. They would give you the P. They would give you the Q. and they would give you all those things, but being able to cross-index, knowing to use that confidence interval formula for that problem would be an entirely – I feel like a different Memory Palace that you need to teletransport to, unless you embed it with in.

Like I said, as I get more advanced, what I'm trying to do, is I'm trying to model what it's like for someone that's going into your methods with fresh eyes and help your viewers figure out this process as someone that's doing this basically from square one. And so yeah, I think as of now I would need to do the cross-index method where you have a different set of rules and a different narrative, different palace, different images, different actions that would be able to reference the narrative of the formulas themselves. As of now, I have not been able to create an expanded version that includes all of them.

Anthony: How long do you think just knowing where you're at now and not knowing how fast you could get and how accurate you could get, knowing what you know about standard exams in this area, how much time do you think you would need to invest in sufficiently memorizing and then rehearsing the material so that you were confident that you had it available as a crib sheet in your mind in order to be successful in a standard exam, an exam of consequence that would make a difference in a student's life to be able to pass it?

Robert: Right. Right. Right. How long? Well I guess that's relative. I think I would say I would tell you that being vaguely familiar with these formulas before, I know that they're – for example I know that there are confidence interval formulas. In fact, I know that there are three confidence interval formulas. I knew that going in, but I didn't know what they were except by just looking. I'd have to look at the papers. It took me 35 or 40 minutes to remember nine formulas. And then with that, I think you can say double that time to get the concepts as well as like what a confidence interval is, what a binomial distribution is, what they're used for and those types of things, which are, honestly, more important than the formulas themselves. You have to be able to define what you're talking about before you take out the tool from the toolbox.

But, I'm telling you 35, 40 minutes for nine formulas is very efficient. And it's not going anywhere especially if now I have the narrative. I can take it with me everywhere. I don't have to consult notes all the time. I don't have to write down again and again. That was my method before, was to write the formula. I said to myself, well, if I don't have the formula memorized, let's see how I feel about that after writing it five times. And then I'll reevaluate whether I've memorized it. So I'd write it down five times. So in answering your question, it's not only a matter of time that's beneficial, but it's the process. It's the process. It's not lame. It's not a waste of time. It's not a drag. It's not a drag on your consciousness. It's more of like a, Where am I? Where am I? It's introspective, which I like. I like that a lot as weird as it was at first.

Anthony: Do you think that there are any students that this method would either not appeal to or just be outright wrong for?

Robert: Yes, I do. I come across a lot of students. I think anyone – and this is not a judgment or any lack of respect for any students or any type of student out there. I think what's happening in education today is that some people really more and more are demanding a spoon feeding system, in which you just say, this is what it is. This is what you need to memorize. Here is what they are. And then they'll go through the process of just trying to pound it into their brain and then once it's done, in survival mode, they'll forget about it and move on. Any student that insists on the results only, that insists on not really focusing on process, that insists on doing it one way that has worked before and not ready to question or adopt new methodologies, this isn't for them. This isn't for them because it seems like you're sort of walking backwards.

You could be spending time working on formulas. You could be spending time making flash cards. You could be spending time doing that stuff that has worked and served you in the past. Why would I want to think about going back in my past and talking to my family at the dining table and doing like this? Napkins, No No No (makes X symbols with his arms). It's very different. And so if you're not ready to step outside your comfort zone and sort of confront a new method that you've never even considered before, then yes, it's better to stick to the flash cards. Or I think you, of all people would not mandate your method to anyone that had sort of a tangible amount of trepidation going in. Like, I don't know, Memory Palace? ... not really for me. You wouldn't want to force it on them. They have to be ready to do the work and go into their mind. Does that make sense?

Anthony: It's certainly voluntary. It's got to be voluntary. You've got to be into it. But I think that a lot of people don't do it because actually I think that the term Memory Palace turns a lot of people off. I think it's an incredibly sexy term...

Robert: I do too.

Anthony: ... but I heard from one guy who said that he couldn't even get started because he disliked the term so much. And he told me – this was an 80-year-old man – he told me that he finally came up with "apartments with compartments." That was his preferred term. I just said, hey man, whatever works. Go with it. Call them "red chickens in a field." Whatever you want. Just get over the

131

hump of what it's called and get over the fact that – it's not that it's just one step back to go into your mind and start doing all this sort of work, but it's really two or three steps back because you need to learn the technique in the first place.

Robert: Absolutely.

Anthony: But it's kind of like being a bit of a person who gets invited to the cockpit. You see all the instrument dials and the plane is already flying. I remember when I was a kid, we were going to Disneyland and this was way back before people were flying planes into buildings and the pilot, or the stewardess came out and said, hey, do you want to come see the cockpit? And I was like, sure! I was like nine years old. And I went up to the front. I was able to actually see the two guys sitting there and everything out in the sky and I just got the sense of wonder of this extremely – it seems extremely complicated.

So at first encountering this it's kind of like visiting the cockpit and you see and you'll say I'll never be a pilot. I'll never know how to do all this stuff. It's just overwhelming. But then, let's just say, you decide to give it a try and before you ever get that plane into the sky, there's a bit of training that's involved. You've got to understand this. You've got to understand that. Oh, there's this principle. There's that principle. But once you get that plane into the sky once or twice, well every time that plane has to go to the pickup thing, it's got to pick up the passengers and it's got to taxi to the runway. But as soon as it gets to the runway, it goes into the sky and it flies and it does it again and again and again. And every time it's successful.

So that's basically what this is. So you first see the wonder. You give it a try. You try to learn the technique and then you learn to fly and you come back and you pick up more and more passengers. The passengers being the memories, the material you've memorized. And then you taxi to the runway and then you get up in the sky. And it's not a bad metaphor as well because eventually the plane is going to land and if you don't keep fueling the jet, the memories are going to fade, but you'll pick up new ones or you'll pick them up again because the same passengers sometimes take the same flight. So I think that's a really nice metaphor that I came up with on the "fly" for how this all works, but you definitely have to step back before you fly every time.

Robert: You definitely have to step back.

Anthony: But, do you think it's a worthy investment? Something that you would gladly teach others to do?

Robert: Yeah. It's definitely a worthy investment. I think it should be an offering because let's talk about the guy that was uncomfortable with [the term] Memory Palace. If someone is uncomfortable with "Memory Palace," then they have one of two choices. They can say "apartments with compartments" and *now* I'm ready to work with you, or they can say, Memory Palace, I'm not into it and they back away. It's all about their attitude. It's all about their attitude. How they approach the thing. Do they use it as an opportunity to find their own in? To give something a try, a reasonable try to take those few steps back and learn the methodology and move forward? Or are they going to look for some scapegoat excuse, I don't like the word "Palace." I'm sure you're going to have people that say – "I don't need a Palace. I'm not royalty. I don't believe in that." And then they'll back away. It's just an excuse to not do it. So I think what you're saying is right, voluntary.

I believe it should be an offering. I'm going to offer it to my students as a methodology, but at the end of the day, it has to come from the learner because it's your journey. It's your images. Even our conversation early on, what we were stumbling on was for you to create your images when it came to these formulas. And then your aha moment was so helpful in that it was really about me creating these learning journeys or this Memory Palace journey because this is my craft. This is what I do for a living. This is what I try to help people do. So I appreciate the fact that you turned it around to the learner. I think that's the same advice that you need to give all your learners is that it's really up to them to try it out and see if it works for them. But they have to take those two steps back and really learn the cockpit, you know?

Anthony: Yeah. I think we should just clarify what you're referring to. We had met before. And the idea of this was that you were going to teach me math or some mathematical principles. I was going to memorize those formulas or principles and then I was going to demonstrate how I did it. And we sort of struggled for about an hour I think. In terms of me thinking, first of all, what are we going to use as an example. Me, not being a mathematician, not having any applicable interest in math as such. I mean we talked about conversion rates and things that have to do with websites and so forth being interesting to me and also I'm just interested in math as a concept, but in terms of the sorts of things that you're presenting now with confidence intervals, this is just ... What can I possibly bring from myself when there's nothing at stake? There's nothing of interest and so forth. So what we ultimately ended up deciding was why don't you learn my method and then you apply it to something that has consequence for you and something that you're clearly passionate about and deeply invested in and interested in?

And I think that really raises a point and I wouldn't want to turn anybody away from any topic at all in education. But I think that it does raise the point that a lot of people are into a subject area because they have an end goal that they aren't really in love with because they have a myth about getting a job on the other side.

So one of the things that might prevent people from having success with mnemonics or memory techniques or Memory Palaces is simply that they're not in love with what they're trying to memorize. They're doing it for some reason that is not authentic. It's not real. It's not love and so that may be the true barrier here because it is going to take an extra investment. Not much. 45 minutes

135

for nine formulas is nothing, you know? It's really, at the end of the day, in comparison with what you can do with that, it may be and it has to be said, that at some level the barrier can be that there's not enough true, authentic interest in the topic in the first place in order to warrant just any kind of learning really. So that may – I just wanted to throw that out there.

Robert: I see it every day. You hit it on the head there. Absolutely. And you know what's interesting is that you might be able to sell, as you were saying earlier today, you might be able to sell the topic through these mnemonics as well. And I'm going to conclude my process of Memory Palaces and this learning technique with you with evidence that I have actually done this subconsciously without even knowing it and it has to do with the quadratic formula. Okay? So look, as someone who's not in an algebra class right now yourself. You have a respect for math as a concept, but there's not need for you to walk around and memorize the quadratic formula. So I get that. So when I try to teach that formula to my students, it's just a bunch of gibberish to them, especially at first. But I came up with the mnemonic that they love and I'm going to share it with you and I've been doing this for years, well before you and I had met. I've been doing this for years and I'm going to share it with you. It may help to have the formula in front of you. Could you maybe write it? I could recite it to you and then you'll look at it with me.

Anthony: Okay. I'm just getting a pen and paper here. My fancy new Collins pen.

Robert: Fancy. Okay. So that's the quadratic formula. Okay? So here's the mnemonic that I did. I would write it on the board so I would say class, here's the story about the negative boy who couldn't decide yes or no, to go to a radical party. But the boy was a square and he missed out on (-) 4 awesome chicks. And the party was all over at 2 a.m. they love it. They love it. They come in the next day talking about it. It just gave it a narrative. The story of the negative boy who missed out on four awesome chicks. They love that stuff. And it's just interesting that we found our way together and I had been doing it without even realizing it.

Anthony: Well my immediate instinct would be and I think it would help people who struggle even with that and I think it would benefit people who don't struggle with that, using a narrative like that is to locate that some place to actually see that in a particular place. So if I was going to work on that, I would see it either where I'm in the room now or I'd pick a specific Memory Palace and locate it somewhere so that I have a place to go when I'm looking for it and then I would want to actually see that and make sure I spend some time exaggerating the imagery and bringing color and action to it. So that would just be my response to that to add more from the kinds of techniques that I put into things.

Is there anything else that we haven't covered that you think should be mentioned?

Robert: I've talked about everything that I wanted to process. You've given me a lot to think about. Is there anything else you wanted to figure out or inquire from me? Do you need anything else from me? Or however else I can help you with this? This has been great.

Anthony: My only thing is I hope it continues so we can help more people actually adopt these techniques and at the very least do things with the sort of formula that you just shared with the negative boy and maybe experiment with people who struggle with even that and see about adding a Memory Palace component to it to give it for what might for some people give it an extra oomph so to speak, or also as a kind of gateway drug, so to speak, to more mnemonics when they see that sort of power coming together.

Robert: Absolutely. Giving it a shell, giving it a framework as opposed to just a simple cutie narrative because that's what I've done up until now. They can envision themselves at the party. They can see who's there. They can ask why or who the four awesome chicks were. All those sort of add-ons to help the image come to life. Anthony, this has been really, really special for me. I think this is a huge opportunity for me to be able to share this conversation with you and I think you're doing really great work and I hope that whoever is watching this and your students understand that it's from someone that works in academics that used this method for the first time today and learned about it only a few days ago, it's really special and you're really on to something and I'm just really happy to be a part of it. So thanks for this opportunity.

Anthony: And thank you for bringing your expertise to the practice.

Robert: You got it.

Last Chance!

I have created FREE Magnetic Memory Method worksheets just for you. These worksheets will help you take the memory improvement lessons you'll learn in this book to the next level. You'll also be given the opportunity to watch the free video course *Memory Palace Mastery*.

In order to download these worksheets and start watching the videos, go now while these materials are still FREE:

http://www.magneticmemorymethod.com/free-magnetic-memory-worksheets/

www.ingramcontent.com/pod-product-compliance
Lightning Source LLC
Chambersburg PA
CBHW020917180526
45163CB00007B/2780